THE ORIGINS OF
LIVING PLANETS

ALEXANDER J. ENDRESS

authorHOUSE®

AuthorHouse™
1663 Liberty Drive
Bloomington, IN 47403
www.authorhouse.com
Phone: 1 (800) 839-8640

© 2019 Alexander J. Endress. All rights reserved.

No part of this book may be reproduced, stored in a retrieval system, or transmitted by any means without the written permission of the author.

Published by AuthorHouse 12/05/2019

ISBN: 978-1-7283-3578-0 (sc)
ISBN: 978-1-7283-3579-7 (e)

Library of Congress Control Number: 2019920108

Print information available on the last page.

Any people depicted in stock imagery provided by Getty Images are models, and such images are being used for illustrative purposes only.
Certain stock imagery © Getty Images.

This book is printed on acid-free paper.

Because of the dynamic nature of the Internet, any web addresses or links contained in this book may have changed since publication and may no longer be valid. The views expressed in this work are solely those of the author and do not necessarily reflect the views of the publisher, and the publisher hereby disclaims any responsibility for them.

INTRODUCTION

Ever wonder whether we are alone in the universe, or are there many other inhabited planets beyond our solar system? Have you ever found yourself staring up at the night sky and imagining if any of those many stars could have life around them? If you ever believed that we are the only ones, then ponder this: There are hundreds of billions of galaxies in the universe, and 200,000,000,000 stars in our own galaxy. Plus, our galaxy, the Milky Way, creates seven new stars each year (nasa.gov Goddard). Moreover, there are eight planets in our solar system with three of them originally in a position to develop life, if other things went right of course. Furthermore, there are moons within our own solar system that could sustain primitive life. With these astronomical figures the possibility of life is distinctly probable. It doesn't necessarily mean that they are visiting us because of vast distances between stars, the limitations in velocity and problems with carrying life supporting material. However, they are probably out there living billions of trillions of lives separated from us by hundreds or thousands of light years. A light year is the distance light travels in one Earth year, and light travels at 3000,000 km per second, so a hundred light years is unlikely for aliens to travel.

It is true that there's a lot we don't know. It is a well-known fact in science that there's a limit to human knowledge, and we will never come close to knowing everything. For instance, the Heisenberg uncertainty principle in physics states that there is a limit to what we can know about the physical properties of a particle. Then there is the observer affect, also in physics, that states that when we observe a particle, we affect it and altering what is there and what we observe. The point I'm trying to make is that there is a limit to what we can know, and not just in physics but in astronomy, evolution, and cosmology (the study of the creation of the universe) as well. For example, we don't know how many stars have planets orbiting around them and will never know. This is because, for one thing, planets are too small to see even with the Hubble Space Telescope and especially with the very much brighter star close by that makes a dim object like a planet difficult to observe.

However, through mostly indirect observation, scientists have been able to identify hundreds of exoplanets, or planets around other stars (Exoplanet). Because of the way they are detected, most of the detected planets are gaseous giants like our Jupiter, or Saturn, but there are some known terrestrial extrasolar planets (Planet Quest). Most of these are too big for habitation, the gravity would be enormous, but there are a few in the right size range for sustaining life. Also, according to Planet Quest, two times the size of Earth is the maximum size for a planet to sustain life. Furthermore, according to Planet

Quest, which is part of NASA, there are calculated to be "10^{10} Earth-like terrestrial planets predicted by simple statistical models existing in our galaxy" alone, which makes for a lot of possibilities. What's more this doesn't account for exoplanets in other galaxies, which makes for astronomical possibilities for life.

Because an exoplanet is so far away and because they pale in brightness to their mother star, directly viewing them is very difficult. So, scientists have come up with some ingenious ways to detect them. One method is by taking advantage of the Doppler Effect. We can observe the Doppler Effect when we are listening to an ambulance passing by. When approaching the sound waves are condensed and when receding the sound waves are stretched out. What this means is that when the ambulance is approaching, the sound is at a higher pitch, or frequency than when receding. This same effect works for the electromagnetic spectrum (light waves) and allows us to detect a foreign planet. When approaching us, the light is blue shifted, when receding the light is red shifted. What I'm saying is that when light approaches us, it is slightly shifted to the blue of the electromagnetic spectrum because blue is a higher frequency and shifted slightly red when receding, red is a lower frequency. When a planet orbits a star, it is also sharing a mass with its star and alternates between red and blue shifts as it orbits its star. When the planet is moving towards us, we will observe a blue shift because the waves are condensed, when moving away, it will create a red shift, and the waves are stretched. If we observe a blue shift, then a red shift alternating, we can be pretty sure there is a planet there orbiting the star.

Anyone can witness the electromagnet effect by looking into a prism where the light is spread apart into separate colors. It can also be witnessed by a rainbow. The electromagnetic spectrum is a correction of colors, and the combination of colors is white light. A prism or rainbow splits the individual spectrums off into the separate color spectrums.

Frequency and wavelength are inverse. The higher the frequency, the closer the wavelength between peaks with shorter wave lengths. When light is approaching, the light gets compressed, and a shorter wavelength with higher frequency is produced. The opposite happens when the light is traveling away.

Another method used to detect planets is pulsar timing. This was the method used to detect the first exoplanet in 1991 (Cornell University). A pulsar is a very old star that emits light in pulsars that can stretch across many light years and is kind of like a lighthouse. These pulses can be precisely timed and creates a way of detecting objects orbiting around them. When an exoplanet is orbiting around a pulsar star the pulses are altered that can make detection possible. However, a planet orbiting a pulsar is probably not inhabitable because pulsars are chaotic and makes for a hostile environment for any planet orbiting it.

An alternative form of detecting an exoplanet is the transit method also called the photometry method. The passage of an exoplanet, and its star is called a transit. This kind of detection occurs when a planet passes between its star and the observer. When the exoplanet passes between us and its star, the light from its star is dimmed just slightly. Most exoplanets have been detected with this method. The size of the exoplanet can then be estimated by the amount of the dimming, with small terrestrial planets only creating a small dimming effect.

An additional method is the gravitational microlensing method. This method is based on Einstein's theory of general relativity and uses complex mathematics that is beyond the scope of this book. The basics, however, are situated on the fact that heavy objects, like a planet, curves the space-time continuum around it. Astronomers look at a suspected planet that pass in front of a background star. The light from the background star then gets magnified in a certain way if there's a planet there.

The Kepler space telescope was launched on March 6, 2009, from Cape Canaveral for the specific purpose of searching for exoplanets. Kepler is the first exoplanet searching satellite (NASA. Planet Hunter). Kepler narrows its search by looking at a small patch of sky with about 15,000 sun-like stars similar in size and type to our own sun in the area of Cygnus lya, (NASA.space.com), utilizing a digital camera to watch for dimming in a star when the exoplanet passes in front of it. As of 2012, Kepler has identified 2,300 possible exoplanets that need to be confirmed by further observations (space.com).

Considering our knowledge of how solar systems are formed and the plentitude of planets, moons and other objects in our own solar system, it is only expected to repeat itself in other solar systems. In other words, the same laws and mechanisms that formed our planets are very plausibly at work in other solar systems. After all, there are certain laws of nature that are universal, such as the laws of physics, evolution, and gravity, so if our solar system developed planets, there is no reason not to suspect that at least some other solar systems developed in a similar manner. This does not mean that all solar systems develop like our own, but only that given the trillions of stars out there, it is expected some have developed planets, and that some of these solar systems have developed in a similar manner as our own, which means that there's plausibly planets that are capable of the conditions needed for life with many teeming with life.

What I'm arguing is that life starts by chance once the conditions are right, and with trillions of stars, there are plenty of chances for the development of life. By chance, I mean out of the trillions of stars only maybe a few trillion of them may develop small terrestrial planets like Earth. Out of these trillions of Earth like planets maybe only a several billion may start the conditions for sustaining life. And out of these billions starts of life, maybe only a few million make it past the very primitive stage, while the rest simply die out like Mars possibly did. Hence, maybe only a few million stars in the universe may develop advanced life forms. Then what's more, these stars are spread out for billions of light years within the universe.

I also argue that once the conditions for life are met and life starts, life develops by the same laws of evolution that occurred on Earth. Only after millions of years of evolution and the right conditions, higher intelligent life capable of language and advancing themselves may develop. As Einstein once said, "God does not play dice with the universe," meaning the development of nature follow laws and are not random or based on probability. Darwin himself stated that life develops by chance with the right conditions, and then precedes by certain laws, and these laws I'm saying are universal laws of nature that duplicates in extrasolar planets capable of supporting life. I can't see any reason why the laws of evolution wouldn't be similar on other inhabitable planets. After all, the theory of evolution is based on natural selection and extinction; the fit survives and pass down their genetic codes while the unfit simply die. This law of natural selection seems only sensible to repeat itself on exoplanets once life starts there. I find evidence for a general law that duplicates itself on our own planet. For example, different species on isolated continents and islands all followed the same laws of evolution, so why not on far away planets?

Another point I would like to make is that language and the cognitive niche is most of what makes possible highly developed life forms. There are other things like hands and walking upright, which helps in structuring advanced species. However, with language, I suppose creatures could build and write with something other than hands, say toes, mouths, or beaks. Yet, without language, ideas cannot be shared, advanced tools cannot be built, and knowledge cannot advance and be passed down through generations. Even the simplest tools take language, enabling people to communicate how to make things and share ideas. Just imagine making a simple ballpoint pen without language. What's more, it takes language to build on previous knowledge to make better tools or ideas. Take the simple ballpoint pen, which didn't come onto the market until 1945 (ideafinder). Before that there was the fountain pen, which was patented by Lewis Waterman in 1884. The point I'm trying to make is that even the simplest tools and ideas takes years of development, which is only capable through language. Language and the cognitive niche go together; I doubt if any creature could have language without higher intelligence, and higher intelligence would be kind of useless without language to pass down knowledge and build on ideas.

The first section of this book is about how living planets develop and evolve and review of the possibilities of ETI (extraterrestrial intelligence). Some of it is analysis of what knowledge we have. For example, I give you what is known about our solar system development, the evidence of other planets and evolution. I give you this review because it fits into my general theory of how life develops, and I provide evidence to support my theory. Then, I talk about life on other planets and the possibility of visitors, which I disagree with. My stance is that although life is most likely plentiful in the universe, it is sparse between life sustaining planet with maybe hundreds or even thousands of light years between intelligent life forms.

This takes us to the second half of the book, which is that planet Earth is unique among planets within several hundred or thousands of light years. Sure, there is probably life on other planets, but intelligent life is most likely hundreds or thousands of light years away, which makes Earth exceptional within a local distance. Hence, because Earth is relatively unique, we should treat it as the exceptional planet it is. This means taking care of the environment, other creatures and controlling our exponential population growth. The course we are on now is destroying our environment: Global warming, population expanding exponentially, toxic waste, never-ending landfills, species extinction, nuclear waste, threat of nuclear war, etc. We are now treating our planet like it is disposable or like it's one of many and can be easily replaced once we destroy it. We need to start taking care of our planet and treat it as exceptional as it is. Therefore, the second half of the book, I will talk about evidence of global warming and other environmental damage then talk about solutions.

A majority of section two then is evidence for global warming and other environmental concerns and what can be done to halt it or at least attenuate it's advance. I will also talk about other threats to our environment, as mentioned in the preceding paragraph. Global warming is one of the largest threats to our planet, our own survival, and national security. There is overwhelming proof for this in the scientific community; however, outside of the scientific community, such as among politicians, there is skepticism. Some politicians believe that global warming is a "hoax perpetuated by the Chinese." In all reasonable thinking, why would someone think the Chinese would do this? A large part of the second section will be going over the proof of global warming in a hope of convincing the skeptical reader of urgent attention. Then, I will take my reader into the possible solutions to these issues.

What is novel about this book, as far as I know, is that I introduce that living planets, with life, develop by chance, then the same laws of evolution that occurred here on Earth take over on these alien planets. Then I argue that Earth is unique within many light years. To do this, I dispute UFOs as being ETI; I use the vastness of space, the speed limit in space, and the difficulties of interstellar space travel to do this. My final and most important argument is that because Earth is exceptional, we ought to treat it as exceptional and preserve our environment, soil, waters and other species.

My idea is simple; if there're enough chances in nature, even the very unlikely is probable. The chance of life in the universe may be as low as only one out of a million, but with trillions of chances there's a very good likelihood that alien life does develop. Many of these life forms may be only primitive, such as insect-like dominated planets, or primitive reptile-like, or wild mammal-like creatures, or primate-like creatures, or most likely something I can't even imagine. However, with chance occurrences, trillion of possibilities and millions of years of evolutions some of the inhabitants of living planets will develop language and a cognitive niche, which leads to higher developments and knowledge in some cases far beyond ours. I also include a chapter on UFOs and related subjects, make some informed speculations about other life forms, and predict possible futures of all life forms may take including our own.

Then the primary point of this book: With many light years between ETI and us, the vastness of space, the speed limit in space and difficulties with interstellar travel, ETI visiting us is unlikely. Hence, I claim that intelligent life on Earth should be preserved and treated as precious and not the way we are doing presently.

PART I

THE UNIVERSE, UFOS, EVOLUTION AND PLANET FORMATION

CHAPTER I

FORMATION OF SOLAR SYSTEMS

Explanation

BECAUSE THIS BOOK IS partly about the origins of living planets, I feel it only proper to start when a solar system first develops, as opposed to the start of life or the Big Bang. This is because it is when solar systems develop is where the chance for life has a possibility for development. Life can't develop in space, and the seeds for life have already taken root once the conditions for life are made possible. These conditions are a rock, or terrestrial planet large enough to obtain sufficient gravity, but not too large to crush its inhabitants with overwhelming gravity, but large enough to hold an atmosphere with enough density to support an atmosphere. The limit in size of a terrestrial planet is considered to be no more than twice that of Earth. It also needs to be the right distance from its sun that has a similar composition as ours. I doubt, for instance, if an x-ray star can support life. Another condition may be a gaseous giant, like Jupiter, in its outer solar system to absorb incoming large objects; otherwise the inner terrestrial planets would be compacted with meteorites. When these conditions are met by chance only, then it makes possible for other necessary items for life, such as water, an oxygenated atmosphere, certain elements and so on.

Solar Systems' Early Histories

Our solar system started out in a giant gas cloud. At some time, it started to condense by some unknown explosion like maybe a nearby supernova. Incidentally it is an exploding supernova that is thought to have created carbon that is the basis of life, so we are literally made of star dust. This explosion started the gas cloud to contract and spin. As it condensed in the center it also got hotter until a sun was finally formed with spinning rock and gas around it. As the cloud thins out particles are attracted together through the laws of gravity, which will eventually form planets, moons and other bodies. Near the center of this swirling dust of gas and rock-solid planets formed (Mercury, Venus, Earth, and Mars). Because solid material is the only thing that could tolerate the intense heat terrestrial planets formed in the inner solar system. In the outer regions, ice and gas could be formed, which eventually became the gaseous giants, Jupiter, Saturn, Uranus and Neptune. Pluto is thought to have been captured from the Kuiper Belt, a system of asteroids beyond Neptune and Pluto is considered a Kuiper Belt object—not a true planet.

As far as distance from the sun is concerned, I think

that three planets could have developed life if everything else happened to go right. These three planets are Venus, Earth, and Mars. Mercury is too small and close to the sun. Unfortunate for Venus, though, it developed a carbon dioxide atmosphere with sulfuric acid and sulfur dioxide clouds. Plus, it has an average temperature of 860 degrees F, not exactly life-building material. Venus is just one way a planet can go to building a non-life sustaining climate. What makes life so amazing is that there are many ways to go wrong, like Venus, but only a series of continuously right circumstances to build life, and it's all by chance and universal laws of nature.

Mars may have started off with a better start, at least at first. There is scientific evidence of nannofossils of maybe ancient Marson micro bacterial life forms. Meteorite Allan Hills 84001 (ALH 84001) was found in Allan Hills Antarctica in 1986. This particular meteorite is one of 12 found on Earth and thought to have originated on Mars, due to their chemical makeup. The nannofossils in question are 20-100 nanometers in diameter and appear to be fossilized bacteria. A nanometer is super small at $1 \times 10-9$ or one billionth of a meter, probably not an intelligent life form, but just the same a life form. Some experts point to the possibility of Earthly contamination of the meteorite. In other words, maybe the bacteria that formed the fossils arrived after it landed on Earth.

Even if these nannofossils came from Earth, which some claim they did, there's still other evidence of the early start of Marson life. Maybe you heard about the recent discovery of the possibility of water on Mars. For example, recent photos of the Marson surface indicates water has flowed there within the last seven years (nasa.gov.press release). Although it is too frigid and the gravity too week on the Marson surface for flowing water, water could flow from underground sources and break free and make water flows with deposits like is reported by NASA. Underground water lakes and rivers are more probable because on the surface water might evaporate because Mars is a small planet and hence has lower gravity than Earth. Where there is water, there is the possibility of life. Therefore, it is possible, even probable, that there is micro life under Mars service in underground lakes that still exists today.

If life started on Mars and still exists today, it indicates that at least the beginnings of life are somewhat common, for to have life to begin on two out of a possibility of three possibly inhabited planets is astonishing. Then figure this into the trillion of possible solar systems out there, the possibility of at least primitive life beginnings is very apparent. However, what about more advanced life than bacteria? What is the likelihood of higher life forms, like mammal-like or human-like creatures or some other form of intelligent life? Well, when you consider that in our own solar system with only three possibly inhabitable planets, concerning their distance from the sun, there's one planet Earth that has plentitude of life. Add to this the astronomical number of stars in the universe, the possibility of life is enormous. One out of two is not bad. Maybe not all solar systems fare as well as ours, maybe some only got as far as Mars might have, or not at all, but sooner or later, there's bound to be solar systems, hundreds, thousands, or millions of light-years apart, that

THE ORIGINS OF LIVING PLANETS

sustain life. I know there must be some skeptics out there. Maybe they think that Earth is unique, and if there is other life, it is inferior to humans. I have several reasons to doubt the accuracy of this idea. First, it is just the mere vastness of space that holds so many possibilities for life. Second, there are certain laws that are universal throughout the universe. Finally, man has a history of thinking he's the center of it all, making up elaborated explanations to support his ideas. What's more, he has been proven wrong in history many times.

For instance, before the early 1600s, most people adhered to the geocentric view of the universe that held that the Earth was the center of it all and everything in the universe evolved around us, literally. There were big problems with this hypothesis, but the ancients came up with elaborate ideas to keep their view. Particularly was the problem of explaining the cycles of the planets. To account for unexpected planetary cycles, the ancients devised epicycles that were superimposed upon circular orbits around the Earth. This hypothesis became increasingly complicated as observations of the planets improved. They needed to keep explaining more and more anomalies to justify their geocentric hypothesis. This is an example on how human in history has clung to a believe that we are special and at the center of everything.

Galileo came up with a much more parsimonies hypothesis. He proposed the heliocentric view around 1610 that held that the sun is the center of the universe. Galileo met with stern opposition of his view. The Catholic Church was especially critical stating that Galileo's view was false and contrary to Scripture, and he was ordered to stop his support for the heliocentric view (quoted in Wikipedia.com). In 1632, he published his most famous work, *Dialogue Concerning the Two Chief World Systems*. Since then he was forced into exile. The point I am trying to make with this example is how firmly man seams to hold on to views that he/she is great in some way and at the center of everything. If we are not the center and special, then the possibility of life elsewhere is probable.

Take Darwin as another example, who to this day encounters religious opposition. I will say more about evolution in another chapter, but I will say that evolution and religion are compatible. You may have to believe in Adam and Eve as a metaphor, but you can still believe in creation. For example, scientists have said that the universe started with the Big Bang, and this could be your creation point. In philosophy, we call this the first mover, but if there's no creation point, then what came before, and before that and so on? It could still be that God, or your favorite deity, maybe Zeus, Poseidon, or other deity or Goddess created the universe at this point and set up natural laws to evolve in the way that I'm suggesting. Some of the stories in the Bible in Genesis could be metaphors. They are still important, but not historical records. Also, I claim that we started by chance occurrence. However, I only make this claim for what happens after the Big Bang, so I won't disagree with you if you believe that the Big Bang was the creation.

Back to the point, though, man has a history and a predisposition to think of him/herself as at the center, or most important, and this may interfere with accepting that we are not the only, or the greatest species in the

THE ORIGINS OF LIVING PLANETS

universe. You may have a problem believing we started by chance and we are not the significant, or highly important one in the universe. Why we tend to think of ourselves as at the center of it all is probably psychologically complex with many variables or contributing factors. One factor is that we are, so it seems, the dominant species on this planet, so why not the universe? However, I suspect that there's a lot more to it than just this. Part of it may be religious; after all, the Bible says that God created man in his own image. This does short of fly in the face of what I am presenting, but then again it could have been meant as a metaphor. Furthermore, there also could be a genetic link. Maybe more confident specimens are more likely to survive because confidence builds on abilities necessary for survival, and then confidence gets passed down to the point where we think that we are the most important in the universe, something I'm arguing is a fallacy.

Evidence

You may have other reasons to reject my theory. Maybe you think that using our solar system to explain others is antidotal evidence or trying to apply the actions and circumstance of one example to generalize to other worlds. Well, I have reasons to suspect that what happened to our solar system has happened to others, other than just the vastness of space. When our solar system developed by a set of physical laws, such as gravity, heat, and evolution, it did so by a set of rules that are present throughout the known universe. Plus, we have evidence that other solar systems started in a gaseous cloud. It was gravity and heat that caused the solid rock planets to form near the sun where incidentally the necessary heat is for making life possible. Gaseous giants that shield the inner planets could form where it was cooler, which would with little doubt be duplicated in other solar systems.

Since we have evidence of large Jupiter-like planets in other solar systems, there is no reason to not suspect that some solid terrestrial planets developed closer to its sun in at least some of these solar systems. According to The Extra-solar Planets Encyclopedia, which I shall call ESPE, there are 347 known extrasolar planets. The site separates the planets into how they were discovered. Most of the discovered were made with radial velocity. In these methods, the gravitational pull of a large planet has on its parent sun causing it to wobble a bit in its orbit that can be picked up with the Doppler Effect.

CHAPTER II

EVOLUTION

Evolution is a Scientific Fact

IT IS ALSO ONE of the soundest theories in the sciences. A scientific theory is a set of explanations arrived at only after numerable years, observations, experimentations, confirmed predictions, and peer reviews that all come to the same conclusion. Examples of theories are gravity, nuclear relativity, and evolution. As contrary to the common use of the word *theory*, a scientific theory is not an idea, or hypothesis, but a scientific theory is the best explanation that we have to explain something. However, evolution is even more than that; it is a scientific fact, and all of the empirical evidence points towards a Darwinian explanation. To support the scientific fact of evolution, we have numerable indisputable evidences that species mutate, or vary in nature and advance or die through a process of natural selection. The better-fit offspring has better chance of survival and passing down its genes because of a mutation that benefited the offspring.

Not all mutations are for the better; some might

worsen the chance for survival while many may not be a benefit or detriment to the creature at all. However, by chance and the laws of nature, some mutations will make the organism very slightly better fit and more likely to pass down its DNA to its offspring. It is important to recognize that evolution is not by chance alone. There are rules for evolution to follow, such as mutation, variability, natural selection, and sexual selection. Therefore, evolution is not entirely by chance, but governed by a set of rules for the laws of chance to follow. Natural selection is the progress of a species to be more fit through mutating and by selecting a more fit form. Sexual selection, on the other hand, is when a mate, usually the female, selects a partner based on a desired trait. A good example of sexual selection is the peacock, where the female selects a male with more colorful plumage making it more likely to duplicate this more colorful plumage.

As in nature, animals of captivity, such as dogs, cats, farm animals, and domestic plants develop through genetics and linage, but the control for what gets passed down are controlled by humans, not nature. This is called artificial selection. The only thing that is different is that animals in nature develop through natural selection in the struggle for survival and to pass down their genes; with domesticated animals and plants, we are the ones manipulating the DNA. For example, I have a Yorkshire Terrier (Buttons), and I know from the literature that they were bred to be small and good mousers. We have done this intentionally by breeding the smallest and the best mousers that pass down the genes for these characteristics to their offspring, and they kept manipulating the DNA for many generations until we got the modern Yorkshire Terrier. Hence, Yorkies are small and good mousers, both of which I can attest to.

What is even more amazing is the modern Yorkshire Terrier, like all domesticated dogs, are the descendents of the gray wolf, which weighs from 55 to 170 pounds depending on location and season. This is a huge difference from the gray wolf and the Yorkshire Terrier that normally weighs around seven pounds or less. How did Yorkshire Terriers get to be so small if they came from the gray wolf, and evolution is such a show process? The answer to this question is simply time, and lots of it, with hundreds of generations. This is a small time in evolutionary terms because artificial evolution works much faster than natural section. The reason being is that with artificial selection, we can manipulate the DNA with a near 100 percent accuracy, while natural selection must rely on chance and the laws of evolution to produce offspring that are often not better fit. Hence, in natural selection we usually are talking about millions of years. However, there is still plenty of time because, according to most estimates, the first life forms on Earth started three and a half billion years ago.

In the natural element, mutations happen by chance and by the environment, as opposed to purposefully by humans. A good example of environment changing how our genes react is skin color. We all came from Africa, which makes all Americans African Americans (we are all the same—not different species), but somewhere along our ancestry, light skinned people, like Asians and Caucasians, lost their dark skin color. This is because darker skin is not needed to protect the body from the sun, and lighter skin allows more light absorption, that is essential for absorbing vitamin D, and for conservation (nature loves to conserve energy and not spend it on things that are not necessary) and because the unnecessary tend to very slowly disappear. The unnecessary tending to disappear from lack of need is a common thing in nature. Therefore, bats go blind in dark caves and flightless birds like the kiwi don't need to fly, and these abilities tend to disappear. Things like sight and flight take valuable energy within the body, so getting rid of unnecessary attributes is beneficial to any creature that don't need them.

Not all offspring have a better chance at survival, we can witness this in that often the smallest of a litter is abandoned or simply can't compete with his/her bigger siblings and dies. However, some will have a very slightly better chance at survival. This could be slightly larger, sharper teeth or eyesight, or many other things. What's more, since the offspring with the better chance of survival is competing for scarce resources with his/her very slightly less fit sibling, cousins, and others of its species, the less fit individuals have less chance for survival and eventually die out, and the very slightly improved members become the new order, and this process keeps on going for millions of years until we have complicated species that we see today. It is important to know that not all of evolution is for the new and improved. For instance, the cockroach has been the same for years and thriving, not at all advancing to higher intelligence. It is because the roach is a master survivor already that it doesn't need to advance.

THE ORIGINS OF LIVING PLANETS

The Family Tree

Because species and subspecies come from earlier species and subspecies, they make kind of a family tree where we can trace back an animal's heritage. For instance, we have traced all domesticated dogs back to the gray wolf thousands of years ago. This is where the dog split from the gray wolf, and if you traced it back further, you will observe more splits from other species earlier. Geneticists call this splitting, when one species breaks off to form a new species, which is not all at once, but a more gradual process. By the way, the gray wolf developed from the Dire Wolf (Canis Dirus), which is how species evolve over time from one species to the next and then the next into a family tree with branches representing species over time. Another example is our own heritage. We went through many predecessors through the years, but the last split from another species was around six million years ago when we split from the chimpanzee (Dawkins, R., & Wong, Y. 2005).

Similar to splitting, when a species splits from a previous species, its most recent common ancestors (MRCA) meets up with a preexisting one back in time. The chimpanzee is ours. Then slowly over time our ancestors kept splitting, forming many ancestors who were not modern humans but weren't chimpanzees either. Some of these pre-modern humans were Momo Erectus (1 million years ago, or 1mya), Homo Hobilis (2mya), Australopithecus afarensis (3.2mya), Homo Erectus (1.9mya), and the Neanderthals, who split off from us (600kya) (Dawkins, et al. 2005; & archaeologyinfo). The chimpanzees also split into Bonobos, and chimps around 2mya. This is speciation, where different species, or subspecies split off into alternate forms from a common ancestor creating new species.

It is important to emphasize that this isn't all at once, at least not in human terms, because there could still be many generations within the time the species split. Each splitting point then is a process of transitioning from the original species to new and improved species for its conditions. As branches spread out on the tree with many splits, separate species and subspecies are formed. A subspecies is a category below the level of a species. For example, the house cat is a subspecies of all wild cats. As species spread out on the tree speciation occurs. Speciation can be termed as the process of biological species formation. An example of a subspecies that split off from humans is the Neanderthal. Speciation is how species branch out on a tree of ancestors. Incidentally, speciation is greater on small isolated island and continents like Australia and Madagascar. This is because isolation from others is one method in which species speciate.

Possibly one of the more famous examples of speciation is Darwin's Finches. While traveling on the HMS Beagle, Darwin made a stop at the Galápagos Islands. There, he found 13 species of finches that were each native to its island. The point is that species and subspecies arrive when geographically isolated from one another. In the case of Darwin's finches, the finches on the separate islands were slightly different from one another mostly in their beak size. Darwin's explanation was that because the finches couldn't interbreed with neighboring finches

THE ORIGINS OF LIVING PLANETS

from other islands, each island of finches developed on their own route of mutations.

However, you may be asking yourself, how did each island get first populated with finches in the first place? And, if they couldn't mate between islands, how could the first ones come from the mainland? One answer is possibly rafting. This is where an animal or two floats across the waters on a raft, or log or something. This may sound easy for travel between islands, but how about from South America, the nearest mainland? The answer is a repeated one: with enough chances, even the unlikely becomes possible. In this case, millions of years gave plenty of chances for animals to raft long distances. And it may only take only one success if a pregnant female, or a female and a male raft and have offspring that can populate the island. Back to the point, this shows how speciation occurs on isolated islands because per area there were more subspecies of a species than on any continent. Below is a drawing showing how species speciate. The chimpanzee would be at the bottom, modern humans at one of the tops and other subspecies in between and the Neanderthal as a species that sprit from human and humanoids.

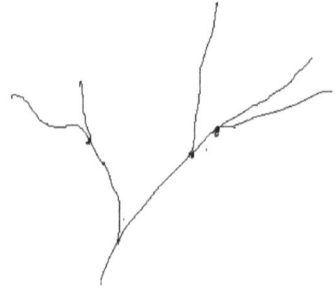

Figure 1: A fictitious tree to represent how species split and speciation is formed.

Sources; my own drawing

Terms Necessary to Understand Evolution

So everyone is clear, and we are all on the same page, there are some technical terms we all need to understand. First, there are different types of speciation. *Allopatric* speciation occurs with isolation of two populations of the same species are isolated due to geographical isolation. A good example is Darwin's finches on the Galapagos Islands where they were isolated from the mainland. Allopatric speciation could be anything geographical including islands, continents, mountains or lakes. Then natural selection and genetic drift (change in the frequency of an existing gene). Genetic drift is also referred to as alletic or Sewell Wright Effect. These terms refer to the change in the frequency and drift and that takes over and new subspecies and with time new species are created.

Genetic drift can be defined as variations in the relative occurrence of different genotypes in a small population by the chance disappearance of certain genes as individual die or don't reproduce.

Then there's *parapatry* speciation that occurs when two species either overlapping zones of habitation or are immediately adjacent to one another. It occurs not when separated physically but separated by extreme differences in habitation. The speciation occurs in the contact zone and presents speciation occurs with variations in mating habits. *Peripatric* speciation is similar to allopatric but occurs when a population is very small. Because there's a small population it may cause what is called the Founder Effect, which is the loss of genetic variation. Peripatry and the founder effect can cause a population to be very different from its parent species. This could happen if the new isolated population carries a rare gene by chance. In just a few generations (if the species are fast breeders with short lives), this new gene could come to fixation, or frequency reaches 100 percent in the new small population.

Finally, there's *sympatric* speciation. This occurs when members of a species develop a new niche that separate them from others of their kind. For example, a new subspecies of the apple maggot fly broke off to form a new subspecies when the apple tree was introduced to North America (Berkeley). Traditionally females only lay eggs on the leaves of the Hawthorne tree, but once apple trees were introduced, some females preferred the traditional course of the Hawthorne, while other females preferred the apple tree. What's more, females prefer to lay eggs in the one she grew up in, and males prefer mates who prefer his same preference he grew up in. Therefore, gene flow (movement of genes through a population) is restricted through apple and Hawthorne dwelling flies. There are lots more to learn about speciation, but this is beyond the scope of this book. If you feel more reading is needed, review to the reading in the bibliography at the end of this book.

Genetic drift is very important to understand but is a complex subject and easily misunderstood. To illustrate, suppose you have a bag full of half white marbles and half purple marbles. Now imagine that when you take 10 marbles out, 100 is replenished at the same white to purple ratio as you took out. For example, suppose you drew out six purple and four whites. The bag would then be replenished with 40 whites and 60 purple. A similar thing happens with genes. Take the gene for brown vs. blue eyes. For many random reasons, genes don't make exact duplicates of itself for the next generation. And it is possible for the genes for blue genes to not be passed down and eventually drift away.

There isn't just one type of evolution, but many types. *Convergent* evolution happens when two or more separate species evolve a similar kind of trait, such as mammals like bats developing wings along with birds. *Divergent* evolution, on the other hand, is when a trait from a common ancestor evolves into separate variations, such as front flippers on sea animals to arms on primates and us (dummies.com). Then there's *parallel* evolution that is easily mistaken with convergent evolution because they both involve the similarity of the descendent. The difference lies in the ascendant, if the descendents were also similar, it is parallel evolution, if not it is convergent evolution. Then there's *coevolution*. This is when interacting species exert pressures on each other, so that they evolve together. For example, bees and the flowers they pollinate is coevolution because they develop together, so of co-developing in a sense.

Then there's a very important aspect of evolution that we haven't discussed much yet. This variable in evolution is *sexual selection*. A good example of sexual selection is the many sexually dimorphic bird species we have throughout the entire planet. Interestingly, attributes that may make the bird easy prey for predators (bright colors) may get passed down through sexual selection because the opposite sex, females, desires it in her mates, which causes the bright colors to be passed down in the males through sexual selection.

Associated with sexual selection is the handicap principal (Wikipedia handicap) proposed in 1975 by Amotz Zahavi. This is because gene flow is not based on longevity, but on passing down genes. It doesn't matter how long the individual bird lives as long as he/she can pass down his/her genetics. Also, of interest is that most dimorphic birds are also polygamous. This is from the Peacock's Tale p. 263: The handicap principle states that reliable signals are expensive to the signaler and costing the signaler something that could be afforded by one with a less particular trait.

ALEXANDER J. ENDRESS

The Beginning

What about that first spark to start the wheels of evolution rolling? In other words, what started life? The study of the start of life, that very first spark, is called abiogenesis, or chemical evolution, but different from evolution. People have been contemplating this field for centuries now. In the sixteenth century, many people believed that life just spontaneously started, which they called the spontaneous generation hypothesis. In about the same time frame, there were people who believed in the heterogeneric hypothesis that states that one form of life comes from another form. They also viewed new data through the foggy lens of their firmly held opinions. For example, when bacteria were discovered in 1674, people assumed it supported spontaneous generation hypothesis because they thought that simple cells weren't complicated enough for sexual reproduction and asexual reproduction through cell division have not yet been discovered. It was in 1668 when Francesco Redi showed how maggots don't form on meat when flies are prevented from landing on it. This was very damaging to the spontaneous generation hypothesis, and the only alternative at the time was biogenesis where life forms come from a previous one. It wasn't until the eighteen hundred that Darwin reasoned that the first spark of life could have started if the conditions are right for life to begin, which then only microorganisms that needed to develop through evolution.

According to this theory, there were three conditions that were necessary for life to begin. First, there had to be molecules capable of reproduction, or copying from a previous form, e.g RNA. Then, there had to be the laws of evolution with duplicates varying from their originals. Finally, these variations had to be inheritable. The one thing we know can copy itself and store data is ribonucleic acid, or RNA. Because RNA can function as a gene and an enzyme, RNA may have preceded DNA and was the start of life (Science Daily, Jan. 10, 2009). This is the RNA world hypothesis. According to the same issue of Science Daily, scientists at Scripps Research Institute have synthesized for the first time RNA that can replicate without the help of a protein or other cellular mechanism. Today, DNA encode genetic information with the help of RNA to carry information from DNA to be decoded into a protein, that are the raw material for chemical reactions, among other things and are considered the "workhorse of the cell" (Berkeley). However, things were different back then, and RNA was capable of replication back then, according to Science Daily. And according to Berkeley, RNA can do much more, such as catalyze a chemical reaction and fuse molecules.

In 1924, Alexander I. Oparin and J.B.S criticized those who attacked the spontaneous generation hypothesis by stating life could have developed spontaneously just once. Oparin wrote about this hypothesis in his book *The Origin of Life*. This spontaneous generation cannot take place now, though, because the environment is not right for it, and any new life form could not compete. Oparin reasoned that life could spontaneous generate if sunlight hits what he called a "primeval soup" (Scribd). According to this theory, about three and a half billion years ago, the Earth's atmosphere mainly consisted of ammonia, methane, and hydrogen with very little or no oxygen. A chemical reaction between these three chemicals occurred with the fuel of the sun, volcanoes, or lightning. The rain would then carry these molecules to a body of water. Where a primeval soup of amino acids, the building blocks of proteins would be formed.

The primeval soup hypothesis was tested and validated in 1953 by two chemists by the name of Stanley Miller and Herald Urey. They used a sealed environment consisting of an assortment of glass tubes and flasks connected in a loop that simulated the Earth's atmosphere three and a half billion years ago. It's important to note that this environment in the tubes was sealed and cut off from the world outside, so that it could not be influenced from the real world. One flask contained water, and another contained electrodes. They heated the water until vapor formed then simulated lightning with sparks from the electrodes. Then they let the mixer cool and repeated this process over and over for a week. After the week was up, they discovered that simple organic compounds and amino acids have developed, where there were none before. This experiment strongly indicates that given the right conditions, our early Earth had, life can spontaneously develop.

Once the first spark of life was spontaneously generated, it started replicating with mutations, some benefiting the organisms and many others not. Reproduction like this creates a family tree with more advanced creatures very slowly evolving from simpler ones. The study of this lineage is phylogeny. Phylogeny does not mean, though, that all species advance from simple to more complex, for many stay simple, and still others die out. Bacteria and other simple organisms didn't evolve much, and the cockroach have been around for millions of years, but still these simple creatures did not evolve to any higher life forms. Take the roach, for example, that formed an important niche, largely devouring decaying food, which facilitates its survival. That is what drives natural selection, survival, and not necessarily higher life forms. And since the roach is a very good survivor, it passed its lineage down as is, and because this early prototype of the roach was better fit for survival than most mutations, the present form was more readily passed down.

Where's the Proof?

If you are a doubtful person who requires hardcore physical evidence to support your beliefs, you may be still doubtful. Or maybe you prefer the biblical explanation for life and are not yet persuaded by my arguments. Well, if you want proof, there's plenty of it. There are fossil records, predictions, similarities between species, birds with genetics for teeth, palimpsests in embryos, bad designs, atavisms, and more.

Let's first start with the fossil records that are very large although not completely conclusive. The only reason I state not completely conclusive is because there are some missing pieces to the puzzle (we don't have fossils for every transitional species), but this is what we should expect if all creatures came about through evolution and were preserved according to physical laws of nature. Bones only fossilize under specific conditions; therefore, what we see through fossils is only a sample of what lived before us. Most bones and other organic matter decompose very quickly. As Darwin put it, "the extreme imperfection of the geological record, combined with the short duration and narrow geographical range of transitional species, made it unlikely that many such fossils would be found" (Darwin, 1999). In other words, because the conditions for fossilization are a rare occurrence and some species were limited in number and range, fossils should then be rare. Despite the rarity of fossilization, we still have numerous fossil records for proving evolution. The reason we still have numerous fossil record despite the rarity of fossilization is the old law of chance, giving enough chances even the rare is likely. In other words, considering the trillions of individual animals that have once roamed this world, numerous fossils are bound to form.

The reason that fossils are rare is because there are only certain conditions that render fossilization possible. For a fossil to form, it must first very soon after death become covered in sediment, which is any matter that can be carried by the flow of water. Slowly over time, the cavities formerly filled with liquid and gas during life are replaced with mineral rich ground water, and the minerals gradually replace the organic gas and liquid. The minerals get into the cell walls where it entirely replaces the former material. The reason that this is very rare is because most organic materials simply rots and never becomes saturated with sediments and minerals.

Fossils have been known about for thousands of years, but only recently has there been a scientific explanation for them. The word fossil comes from the

THE ORIGINS OF LIVING PLANETS

Latin word *fossus,* literally meaning dug up. Many ancient explanations for fossils were based on superstitions. For example, in prehistoric China, fossils were believed to have come from dragon bones and were used in medicine and aphrodisiacs (Wikipedia). Aristotle had quite a bit more scientific explanation for fossils. He observed that fossils from seashells were similar to actual seashells found on the beach and reasoned that fossils must come from formerly living creatures. Leonardo da Vinci also believed that fossils were once living creatures. William Smith noticed in the nineteenth century that fossils vary in accordance with the laws of superposition, or newer material on top of order material. It wasn't until Darwin laid out the foundation for evolutionary theory that a more comprehensive explanation for fossils was given.

Today, we know that fossils are remnants of mostly bones from creatures from the far past. We know their age accurately with plus or minus only about 40 years out of 60,000 years old, which the carbon 14 dating can date back to, and beyond that, there's other methods of dating. Carbon 14 dating was first theorized in 1934 by Kranz Furie, and the first dating came in 1940 by professors Martin Kamen and Sam Ruben at the University of California. Carbon 14 is one of three isotopes of Carbon with 14 being the rarest with only one part per trillion in the atmosphere. Carbon 14 decays very slowly with a half-life of 5,730 years. All creatures have trace amounts of carbon 14, and after the creature dies, its carbon 14 decays at a definite rate, which is carbon's 14 half-life. It is this steady decay of carbon 14 that scientists measure in determining a fossil's age.

ALEXANDER J. ENDRESS

Evolution and Creationism

Evolution does not contradict with creationism religious beliefs. It may take a few modifications, but you can definitely keep believing in creation and believe in genesis. Creation could have come at the moment of the Big Bang, when God set the laws of the universe, such as the laws of physics and evolution in motion. You could think of these laws as being so essential, complex, and universal that some intelligent entity must have set them to work throughout the entire universe.

You can't, however, take everything in the Bible literally and scientifically. This is because much of the Bible is made up of fables meant to set an example, or teach morals and values, or relate with issues of their time. If you take the Bible literally, you will find yourself believing in some unscientific things. For example, the Bible states that the universe is only six thousand years old. However, according to NASA, the Earth itself is about four and a half billion years old, and there's overwhelming evidence to support this. A difference of six thousand years to billions is a large amount. Moreover, scientists have evidence for the Earth's age while religion requires you to believe on faith alone. Faith can be defined as beliefs in something without any evidence; science on the other hand requires empirical evidence, experimentations, observations, and the ability to make predictions and peer reviews. There're hardly any similarities between faith and science in the manifestation of requiring knowledge.

The point to make concerning the development of life on alien planets is that life develops by chance and

THE ORIGINS OF LIVING PLANETS

progresses according to the laws of nature, and these laws are universal. Why wouldn't the laws that drive evolution on Earth not hold for other worlds? Take natural selection for instance where species by chance carry on their mutations. I see no reason why this would not repeat on other worlds. Mutations are mutations whether on Earth or on an exoplanet, and there's no reason to doubt that inheritance is not involved on exoplanets. Once this speciation starts on exoplanets, it must continue just like on Earth. Therefore, there's no reason to question that life starts by chance on exoplanet and continues by the same laws as are on Earth. What this means is that there probably life on exoplanets.

CHAPTER III

SPACE

Barriers to Extrasolar Space Travel

SPACE IS VERY IMMENSELY enormous, velocity is limited within the universe, space travel is bulgy because aliens would have to carry all life supporting material, such as fuel, oxygen, and food, and aliens probably have limited years to live, just like humans. These barriers make extrasolar space travel with beings on board extremely difficult and unlikely. Einstein's theory of E=MC2 means that the speed limit in the universe is the speed of light. Moreover, to get that fast would take an infinite amount of energy. E=energy, M=mass, C2=the speed of light squared. This means that as one approaches the speed of light the energy would have to increase to an infinite amount. Even getting close to the speed of light would be next to impossible because of the enormous energy required. To get a grasp on the enormous obstacles to getting even close to the speed of light: light travels at 70,616,629 mph while the space shuttle traveled at only 17,500 mph. Therefore, us and aliens are limited to only

THE ORIGINS OF LIVING PLANETS

a fraction of the speed of light in space travel, and then only with advanced technologies much greater than ours could any creature approach to a medium fraction the light speed, let's say 25 to 50 percent.

More obstacles to space travel would be the necessity of carrying life support: food, oxygen, a waste disposal system, and the need for gravity. Unless aliens have figured out a way to convert CO to oxygen, they would need to carry bulky containers of compressed oxygen that would take up space and add weight making for the need for a larger spaceship making for the need for even greater energy to get to the necessary speed for sufficient space travel. Plus, there is the need for food and water, adding to the size of the ship and greater energy. Considering all these obstacles, I find interstellar travel by living beings highly difficult.

Some people claim that aliens or us could use "wormholes" in space to instantly travel though the cosmos from point to point. As far as I know, this is science fiction, not science. Maybe in movies, one can use these, but there is no scientific evidence for them. Some tried to use Einstein's general theory of relativity to support wormholes, but there is still no real evidence: no one has ever seen a wormhole, discovered one, traveled through one or used scientific experiments to support them. Moreover, even if wormholes exist in space, we don't know if one could survive by traveling through one without being crushed or distorted in some way that would terminate their lives.

Another option for interstellar space travel could be drones. This would eliminate the need for life support, gravity, and the limitations of space travelers' life spans. I doubt if ETIs could live much beyond 100 Earth years, even with advanced medicines. Plus, I question if alien creatures would want to live most of their lives in space travel. I also question if they would want to die in space travel. This is because, even if you could get here, there's probably not a life span great enough for the return trip.

Drones would eliminate these problems, leaving many generations for travel. However, the problems with carrying fuel and the speed limit is still problematic. Suppose their star is 100 light years distance and with advance technologies, they can reach 33 percent the speed of light, this would give them approximately 330 years to reach us. Moreover, to get signals back to their planet would take another 100 years; radio and microwave signals travel at the speed of light. Yet another issue with drones is that they would need to be entirely preprogramed because to control them by remote control would require 200 years between command: 100 years for signals to reach them and another 100 years to send signals back.

However, if aliens are visiting us, which I strongly doubt, drones would have to be the only option. Moreover, because UFO sightings were often witnessed making fast and quick turns, this could explain things. Humans and probably aliens as well can only stand for a certain amount of G-force. G-force, gravitational of acceleration force, is like gravity with 1-G being the amount of gravity on Earth but results to the force of acceleration puts on the body. Humans can only stand up to about 5 Gs before death. This could make for fast changes in directions, what the witnesses witnessed. However, there

THE ORIGINS OF LIVING PLANETS

are still other obstacles, distance between excepting and receiving signals, and they would need be very advanced technologies to be entirely preprogramed. The advance of drones would be that it wouldn't require life sustaining material, I supposing aliens breathe oxygen, and the limitation of life among any creature who travels vast distances with the speed limit of the speed of light.

To get a hold of the size of what we know of the universe, our closest star (Alpha Centauri) is 4.37 light years away, but it is unlikely to sustain life because Alpha Centaur is a trinary, meaning it is three stars orbiting around one another. The next closest star is Barnard's Star at approximately 5.86 light years from Earth. This star too is unlikely to sustain life because it is a dwarf star and likely too dim to sustain life. The third closest star is Wolf 359 at about 7.78 light years from Earth and is also a dwarf and not likely to sustain life. The fourth star from Earth is Lalande 21185 and located in the constellation Ursa Major, about 8.29 light years away and is also a dwarf. Then the fifth nearest star is Sirlus, the brightest star in our night sky. It is a binary consisting of Sirlus A and B with Sirlus B being a dwarf.

Sirlus is the closest star to us that could possibly sustain life. It is located in the constellation Canis Major. However, it may be able to sustain life, the rule of chance may make it unlikely. Remember, life happens by chance, and it take many chances for life to actually occur. However, even if life did develop there and it is sustainable life, these aliens would have to travel at nearly the speed of light for almost nine years to get to us. Life needs to be sustained in their spaceships with oxygen, food, water, fuel, gravity, etc. Can they carry enough of all this to sustain life for nine years or more like at least 18 years because of the energy needed to approach light speed? My answer is no. This is because of the space and weight these items take up and the more space and weight the more thrust their engines will need. Plus, according to Einstein's theory E=MC2 to get even close to the speed of light would take an infinite energy. So, suppose they can reach fifty percent the speed of light, they would need to travel for almost 18 years while carrying enough life sustaining material with them. This I find astronomically unlikely.

Our own galaxy is 100,000 light years across. Our closest galaxy within the local system (a group of about 30 galaxies including our own) is Andromeda at an approximate distance of two million light years. An example of a galaxy outside our own local group is NGC 3621 at 22 million light years away (nasa.gov). The most distant galaxy detected in the visible universe are approximately 9 billion light-years away and 90 percent of galaxies are yet to be discovered.

But what if intelligent life only develops around let's say one in ten life sustainable stars, and then only a small fraction out of these may advance to higher life forms capable of space travel at such far distances? Remember, that life starts by chance and develops by the laws of evolution, so not every star is going to have intelligent life around it and every ten is likely a liberal estimate. Looking at it from this perspective, this means intelligent life far advanced from us (and capable of far space travel) is probably only one in maybe a thousand stars or more. This would put the nearest star of very advance species to be at least hundreds of light years away. Another thing we need to consider is that Alpha Centauri is a trinary system, and we don't know if life can develop around three stars. Considering the likely distance, unless they found a way to break the light barrier or found ways to advance life to hundreds of Earth years, they would have great obstacles to such space travel.

More chances other than Mars that we have already discussed are the moons of Jupiter and Saturn. One possibility is the Jupiter moon Europa that is thought to have an underground sea beneath 10 to 15 miles of ice (nasa.gov.europa). And the ocean is estimated to be 40 to 100 miles deep. Europa that is named after the daughter of Agenor in ancient Greek mythology is thought to possibly harbor primitive life below its surface. Galileo made numerous flybys of Europa between 1998 to 2003. What Galileo have found is odd domes that suggest that the moon's surface is slowly turning over or convection possibly from heat rising below the surface, and heat and an ocean could theoretically mean there's life below. The Europa Chipper a future spacecraft planned for a later time will have radar to penetrate the surface to determine if there is an ocean below. If there is an ocean and heat the possibility for life is good (NASA solar system/moons/Saturn).

Yet another possibility for life in our own solar system is Titan, a moon around Saturn. This is Saturn's largest moon and second largest moon within our solar system (nasa.gove.titan). The Cassini spaceship that orbited and made gravity measurements and revealed, like Europa, the moon has an enormous underground ocean of liquid water likely mixed with salts and ammonia. The European spacecraft Huygens also measured signals during the spaceship's descent into the moon's atmosphere. The spacecraft's measurements strongly suggest a large ocean 35 to 50 miles below Titan's surface. In addition, Titan's surface lakes and rivers of methane and ethane further suggest the possibility of life on this moon. Again, where there is water the possibility of life exists (NASA solar system. gov/moons/Saturn).

CHAPTER IV

UFOS

Pre-history of UFOs, Unidentified Flying Objects

ALL THAT UFOs MEAN is that they are unidentified and flying objects. It doesn't necessarily mean they are from aliens, even though many equate them with extraterrestrial life. No one knows for sure how far back UFO sightings go, but some have claimed they go back to ancient times (crystalinks). The first recorded ones go back to Egyptian hierographic and paintings on caves. Maybe the first recorded, or at least one of the first, is in the notes of a great Egyptian pharaoh, Thutmose III, who reigned from 1504 to 1450 B.C. (Rense). Thutmose III reigned for far more years than the average pharaoh and is often known as the "the Napoleon of Egypt" for his military conquers (J.H. Breasted). Translated from hieroglyphics, he claims he saw flying discs and fire coming out of the sky, which he described as a "circle of fire" (anomalies.net). Then the aliens took him aboard one of their spacecrafts, and he "flew up to the sky" and then "learned the mysterious ways of heaven" (anomalies.net).

Alexander the Great even reported an UFO that dived at his horses while he and his men were attempting a river crossing. He described them as "gleaming silver shields" that steeped down and scared his horses, causing them to stampede (wipnet). Then it came back seven years later, and this time sent out a beam of light that destroyed some of the wall of the city of Tyre (physics forums). Who knows if any of these accounts are of actual sighting or just stories, but regardless, they certainly indicate that UFOs are not a modern phenomenon.

Modern History of UFOs and Scientific Study

The first "scientific study," that I'm aware of, was conducted under the U.S. Air Force with Project SIGN, started in 1947 to study several recent reports that occurred of UFOs. SIGN was formed after Brigadier General George Schulgen did a study of the most famous sighting with reference to public attention at the time. The general determined that the sightings were real crafts and not fictitious. This most prominent sighting occurred earlier in 1947 by a businessman and private small aircraft pilot who claimed he saw something strange and unexplainable in the sky. While flying over Mount Rainer in Washington on June 24, 1947, Kenneth Arnold described seeing nine elongated saucer-like objects flying a circle pattern around Mt. Rainer.

His original flight plan was from Chehalis, Washington, to Yakima, Washington, but got rerouted to search around the southwest side of Mt. Rainer for a large transport plane that supposedly went down there

(Project 1947). When flying near Mt. Rainer at about 9,200 feet, Mr. Arnold was startled by a "bright flash" that reflected off his plane (Arnold, K). This flash startled him because he at first thought that he was too close to another aircraft. He looked all over but couldn't detect where the light came from. This is, until he looked to the north of Mt. Rainer, where he observed what he thought at the time were nine strange looking aircrafts approaching Mt. Rainer rapidly at around 9,200 feet. A few of them were changing altitude (as much as a thousand feet) and course that made them reflect light, which explained the first sighting of a bright light. Although he assumed they must be a jet aircraft because of their speed, he could not detect any exhaust on any of them. He estimated his distance from them as from 20 to 25 miles and that they must be very large to make out their shape (that he described as "saucer like") at this distance (Arnold, K).

The media have accused Mr. Arnold of seeing reflections or a mirage (Arnold, K). To this, Mr. Arnold answered that this was untrue and that he had an unobstructed view out his open side window for a rather long time. Incidentally, this is when the press first coined the term "flying saucer" (Wikipedia.org). When asked by many people to take his best guess as to what the objects could have been, Mr. Arnold replied that "it is just as much a mystery to me as it is to the rest of the world" (Arnold, K).

SIGN, which was the objective to study these cases, was taken very seriously by the Air Force who assigned it an A2 rating with A1 being the most important. Astronomers were used to weed out astronomical occurrences, such as meteorites, or another phenomenon. Aeronautical engineers, missile experts, nuclear physicists, and intelligence officers were also assigned to SIGN. The era had a lot to do with the concerns over UFO sighting because the 40s were the height of the cold war, and there were concerns that some of these UFOs were Russian. Our government was alarmed over national security either from the Russians or extraterrestrials.

Maybe SIGN's first investigations—and most drastic, were a sighting then known of as the "Mantell Incident," named after Captain Thomas F. Mantell who died "while in aerial pursuit of a UFO" (encyclopedia of science). On January 7, 1948, there were several witnesses who claimed to have seen large oval objects. One witness was Sgt. Quinton Blackwell, who reported seeing strange objects in the sky from the Fort Knox airfield control tower. Also present was Colonel Guy Hix, who said unidentifiable objects were seen about a quarter size of a full moon and turned back after all three were ordered to back off, but Capt. Mantell refused and approached the objects with his single engine, one-seater prop fighter aircraft; a P-51 Mustang that first entered service during World War II. Capt. Mantell was an experienced Air Force pilot with over 2,000 hours of flight time, so it's difficult to imagine how he made this mistake, but SIGN determined that Capt. Mantell passed out because he flew too high (over 20,000 feet) and didn't have oxygen and crashed. Before Capt. Mantell lost contact, he made some confusing reports (because of static in the reception) of seeing a tremendously large metallic objects, but the tower personnel couldn't be sure what he said (Wikipedia). His plane was last seen in the air erratically swirling (obviously out of control like he was unconscious) down to Earth.

There were a few explanations that SIGN have come up with. One was the so-called Venus theory (or more correctly hypothesis) that holds that Capt. Mantell (an experienced pilot) mistaken Venus for an UFO, but this hypothesis was given up because it was later determined that Venus would have been invisible to Capt. Mantell at the time. A perhaps more probable explanation is a military balloon hypothesis. At the time, the U.S. Air Force had a secret mission (unknown to Capt. Mantell, the control tower and the other two pilots) using large balloons for scientific purposes. Although a balloon couldn't be positively identified as being in the area, the air force did launch several balloons just 150 miles away earlier in the day.

It certainly possible that some of these balloons have drifted off track, and this was what Capt. Mantell saw. Also of importance is that the balloons matched Capt. Mantell's explanation of what he saw, large metal objects. These balloons were made with reflective aluminum and were about 100 feet in diameter. Since Capt. Mandall didn't know of the balloons, he could have mistaken them for something unidentified, or a UFO. What's more, although it seems far-fetched that an experienced pilot like Capt. Mantell could have made such a mistake and then climbed too high for adequate oxygen, this and the extraterrestrial hypothesis seems an apparently viable explanation. However, this didn't stop the media from getting a hold of the story and print fantastic claims, such as Capt. Mantell was shot down by an alien spacecraft, and his body was found riddled with holes (Wikipedia Mantell UFO incident).

To add even more mystery to the case, according to the National Investigations Committee on Aerial Phenomenon, or NICAP, there is evidence that contradicts the balloon hypothesis, in explaining Capt. Mantell's sighting. First, there were witnesses who claimed to see up close something far different than a large balloon. Kentucky State Police and other witnesses on the ground claimed to have seen an object 250 to 300 feet around moving at a very fast speed. As far as I know, balloons don't move very fast, or at least not faster than the wind can carry them. Another problem with the balloon hypothesis is the balloons were witnessed by several reliable observers, including astronomer Dr. Carl Seyfert (who incidentally discovered the Seyfert galaxies), several miles to the north at approximately the same time as the UFOs (NICAP). How could some of the balloons drift off while the rest went far north when you consider that they were all dependent on the same wind direction and velocity? This adds doubt to the balloon hypothesis, and I don't know of any other explanation other than the farfetched theory of extraterrestrial intelligence, or ETI. However, just because we can't explain them doesn't mean we have to accept non-Earthly explanations. Perhaps Capt. Mentells' lack of oxygen, which eventually caused him to go unconscious, also caused him to hallucinate. I find this to be the most likely explanation. The lack of oxygen could theoretically have caused Capt. Mantell to mistake shiny balloons for flying s saucers.

Another interesting case at the time and is possibly

THE ORIGINS OF LIVING PLANETS

the best known today happened in Roswell New Mexico on July of 1947 when Mac Brazel, a rancher outside of Roswell, discovered pieces of unknown metal scattered throughout the Foster homestead where he was a Forman (Center for UFOs). According to many witnesses, there were recovered space craft along with dead alien bodies. What really happened at Roswell, even though the government denies it. Mr. Brazel took some pieces to experts in Roswell Air Force Base for identification and were evaluated by intelligence officers Major Jesse Marcel and Captain Sheridan Cavitt. Initially the Roswell Army Air Field released a statement that they found a "flying disc" on the ranch and then claimed it landed on the ranch, which created widespread rumors of space aliens and such. However, the statement was retracted within an hour and the samples were then ordered to Ft. Worth Army Air Field for further examining. The Ft. Worth personnel determined that the metal was from a weather balloon lined in reflective tinfoil.

The issue seemed to be solved and forgotten until 1978 when the worker who accompanied the findings to Ft. Worth was interviewed and said that he seen something different than what Ft. Worth had said. Also, Major Mercel said that the material was "not of this world," and said that what was shown to the public by Ft. Worth was actually a replacement from what was really found on the farm (quoted in Roswell). Mercel continued describing other materials, such as small beams with hieroglyphics that looked something like balsa wood but wasn't wood and wouldn't burn either. Mac Brazel, the rancher who discovered the debris described it in 1947 as tinfoil, rubber, tough paper, and sticks.

I'm having trouble taking Mercel at his word. First, his testimony came years after the fact. If he found the material to be out of this word, then why didn't he come forward earlier? He then stated that the material found was replacements from the actual materials. Then he makes fantastic claims of hieroglyphics. Was he stating that this was language from an alien lifeform or from the Egyptians? I'm having the tendency that Mercel's claim is either fictitious or imaginary or both if that is possible. It seems that UFOs bring out the wild imaginations of people or even entices them to exaggerate or even make things up, and I feel that this is what happened with Mercel. Plus, memories fade after many years that makes for distorted recounts.

Family members gave more detailed descriptions. His daughter who was present at the time of the finding described something like aluminum tinfoil with some having lettering on it but no words. She also stated that some of the metal had tape on them that wouldn't peel off and when held to the light you could see what looked like "pastel flowers" (Wikipedia Mantell UFO incident). Marcel's son Jesse Jr. also said his father brought home what looked like hieroglyphics on it, backing up his sister's story. First Lt. Robert Shirkey the base assistant operations office who witnessed the plane to Ft. Worth being loaded also reported seeing hieroglyphics on some different looking metal pieces. And more witnesses gave testimony to witnessing similar things as the above.

Another difficult case to solve and one that would

THE ORIGINS OF LIVING PLANETS

eventually determine the fate of SIGN over conflict between Project SIGN and the air force's opinion about SIGN's conclusion. SIGN was the first secret U.S. Air Force project to investigate UFOs. The case in question is the so-called "Chiles-Whitted Encounter" (UFO Casebook, & Project 1947) In the Fall of 1948 piloting their commercial DC 3 Captain Clarence Chiles and his co-Captain John Whitted witnessed to what they called a cigar or torpedo like object with an orange exhaust and with rows of windows that looked like passenger windows and was approaching them at such a fast rate and nearness that the captain did quick maneuvers to avoid the UFO that got within a thousand feet from their aircraft. Also, there was one passenger, Clarence L. McKelvie, who claims he saw a very bright light running parallel to the plane through his passenger window. Yet another corroborating witness was Walter Massey the air force ground chief at Robins Air Force Base about 150 miles away who witnessed a cylindrically shaped object about an hour before Chiles and Witted seen their sighting. This is four reliable witnesses and in three different positions that reported a similar phenomenon.

What could possibly be the explanation for these sightings? A weather balloon, blimp, or some other manmade aerial object of this shape couldn't possibly move at the rate of speed the witnesses have reported. SIGN claimed that the only logical explanation for the sighting was extraterrestrial, but the air force strongly disagreed with SIGN's conclusion, which ultimately resulted in the termination of SIGN. SIGN's report of the situation that supports the extraterrestrial theory was destroyed by the air force and tried to cover it up from the public.

To replace SIGN, the air force created Project Grudge in 1949. Unlike SIGN, though, Grudge operated on the assumption that all UFOs can be explained by natural phenomenon. In fact, according to Edward J. Ruppelt, Grudge took the UFO as being quite impossible and rather silly and everything they investigated was seen through the foggy lens of bias. Grudge's final report stated in a nutshell that UFOs can be explained though natural phenomenon, such as meteorites or the sun; conventional aircraft; war fears, or your typical mentally disturbed witnesses. And possibly most important to the air force Grudge concluded that UFOs pose no threat to national security (Wikipedia project grudge). Probably Grudge's biggest case and ultimately its downfall was a UFO sighting that become known as the Mount Monmouth Case. On a late morning in September of 1951, the crew at a radar facility in Mt. Monmouth New Jersey spotted a very fast and low flying UFO in their radar screen. The object was moving at about 700 mph that was faster than any conventional aircraft (outside of experimental aircraft).

The crew's radar facility lost the object's radar signal around Sandy Hook, NJ, just south of Jersey City, NJ. But just about 17 minutes later just south of Sandy Hook pilot Lieutenant Wilbert S. Rogers and his co-pilot Major Edward Ballard Jr. They observed the object make aeronautical maneuvers that was apparently intelligent driven. They also estimated its speed at 700 mph and got a good enough view to estimate its diameter at 30 to 50 feet and "disk-like" (BB Archives). After the UFO completed a 90 degree turn, it went out to sea, and Rogers could not keep up pace with it with his jet. Later, he described the object as something he never saw before and definitely not a balloon because balloons don't move at near the speed of sound, at 700 mph.

But this is what Grudge has determined that Lieutenant Rogers and two radar sightings detected, with the other radar sightings being natural phenomenon (bluebookarchives.org) based on two balloons being launched in the area. Grudge states that the object appeared bigger only because it was closer than the pilot believed, and the speed was explained with an illusion based on the time elapsed between the first sighting and after he started tracking it that made it appear to be moving much faster than it appeared (bluebookachives.org). Concerning altitude, the balloons were reported at 18,000 feet while the radar and Rogers reported the craft at around 5,000 feet. There were some problems with how Grudge have collected and interpreted data resulting in "sloppy debunking" that bothered General Charles P. Cabell chief of Air Force intelligence. When General Cabell learned of Grudges bias while stating they vigorously instigate every claim, he dismantled Grudge and ordered Project Blue Book in 1952.

Project Blue Book (BB) was created to replace Grudge and Captain Edward J. Ruppelt, an experienced and decorated airman from WWII and with an aeronautics degree would shortly become the first director of Blue Book. Blue Book, the longest running UFO investigation and possibly the most heard of, was a U.S. Air Force project from 1952 till December 17, 1969. BB received 12,619 reports and over 700 were classified as unidentified. The air force conducted these investigations to determine whether UFOs are a threat to national security, which sounds a little paranoid, and to simply explore what could be gathered on UFOs.

This first case that Blue Book investigated occurred in Lubbock Texas on August 25, 1951 (BB Archives). It started when four Texas Technical College professors were observing meteorites that were part of a study when around 9:00 in the evening they noticed twenty to thirty lights in the night sky. They described them as bright as a star but larger and traveling at a high velocity. An astronomer was present who tried but failed to measure the lights altitude. Over the course of a few days more, sightings were made with a total of 12 sightings.

Then on the 31st of August, a college Texas Tech. freshman saw the lights and photographed them. There was one thing different in the photo that wasn't spotted before; the lights in the student's photo appeared in a V-shaped pattern, which caused some doubt to the authenticity of the photo. That maybe it was a fake. Another hypothesis is that the lights were birds with reflective underbellies. The game warden, however, doubted the bird hypothesis, but stated if they were birds, they would be in the plover family because these birds have white underbellies that could reflect light. Yet he questioned this because he doubted whether there would be that many in the area. Incidentally, another sighting of V-shaped lighting was sighted in Bethesda Maryland on the 18th of April in 1952. Four civilian witnesses claimed the lights were "orange-yellow" (BB Archives). BB reported that there was no aircraft in the area and gave no conclusion, another true UFO's (BBA #8).

Yet another unsolved UFO happened on the 12th of February 1952 near Washington DC. First two pilots in their fighter aircraft spotted a vivid white object about one-sixth the size of a rising moon and eight to 10 miles away. The second time they saw the object, it was on a course to Washington DC, but then accelerated and turned south of DC, then hovered for a few minutes and finally accelerated and disappeared. The only solution for this case given was possibly a helicopter because it resembled a previous case involving a helicopter, but no croppers were reported in the area and the case is unsolved.

Another interesting case happened on the 24th of February 1952 by a navigator of a B-29, which was a heavy four-engine prop bomber used during WWII and the Korean Wars. This happened in South Korea at 11:15 Korean time and involves one witness. The navigator described the object as a cylinder and with an exhaust three times the object's length and both the UFO and its exhaust a bluish color (BB Archives). The object then turned towards the aircraft and takes on an interception course and continued for about 15 seconds. Then it descended under the aircraft while the exhaust got smaller. The navigator described it as car-sized from 3,000 feet. Despite this remarkable report from a seemingly credible witness, Blue Book never gave a definite conclusion. The BB Archives pretty much states that the UFO resembled an air-to-air missile.

An unsolved case involving radar and a chase with

aircraft happened on the 29th day of July, 1952, in Michigan. An F-94B aircraft was diverted from its practice flight to investigate. The radar station requested a visual on the object, and when the F-94B got close enough and in position the pilot noticed "a bright, flashing, colored light" (BB Archives). At first, BB officers thought it could have been a star, but that was quickly rejected due to the radar signal. Also rejected was a balloon hypothesis because the reported speed was much too fast.

Another USAF chase with a UFO happened in Los Alamos, New Mexico, on July 29, 1952. Pilots and other witnesses claimed to have seen a "shiny metallic color" object. When first spotted by the pilot, the UFO disappeared but quickly reappeared in front of the aircrafts. Then the UFO made a 360 degree turn to circle behind the fighter where it remained for a couple of minutes then disappeared again. One discrepancy the BB pointed out is that the pilot reported seeing the UFO for 30 minutes and moving at a high speed. According to the BB Archives, to stay in the pilot's view for that long the UFO would have to be traveling at low velocity. The BB classified this case as "unknown."

Perhaps Blue Book's most renowned, difficult to solve and bazaar case was another UFO case over Washington DC. This one in the summer of 1952. It started around 20 minutes to midnight on the evening on the 19th of July. At Washington National Airport, air traffic controller Ed Nugent took temporary controls of the radar panel for his boss Harry Barnes (nicap.org). Shortly, Nugent picked up seven returns on the radar. This was a slow night so far, and there weren't any aircraft known to be in the area. The images were odd from the beginning, and the UFO wasn't following any flight path. Nugent called for Barnes to look, and Barnes later noted that they moved completely radically compared to ordinary aircraft (Clark, J.). Barnes then radioed the other tower at the airport to check on the object. The controller there, Howard Cocklin, not only had them on his radar but had a visual on one of them out the tower's window.

When one UFO hovered over the capital and two over the White house, Barnes notified Andrews AFB, and Andrews confirmed the UFOs on their own radar. Barnes requested jets to investigate, but the fastest Andrews could do was 30 minutes because their runway was under repairs and the jets were temporarily in Delaware (http://www.nicap.org/wns.htm). Andrews and National tracked the UFOs for several minutes Jim Ritchey, another controller and witness, first noticed that one of the UFOs were rapidly approaching an airliner that just took off from National, but then came to an abrupt stop. The pilot of the airliner, S.C. Pierman, said he saw one, but then it accelerated and went out of sight within three to five seconds. And it also accelerated out in a couple of blimps of the radar screen, which enabled the controller to measure its acceleration. They estimated that it went from 130 mph to 500 mph in three to five seconds (NICAP). Then one of the UFOs took a rapid 90-degree turn, something that humans can't do (nicap.org). When the sweep on the radar came around again, another UFO suddenly stopped from 100 mph reversed and reappeared on top of the other blimp all within five seconds.

On another radar, Joe Zacko picked up an object

that was moving at the astonishing speed of 7,200 mph (NICAP). First it descended from a high altitude into the Zacko's radar, leveled off for a few seconds and then climbed back up again very fast out of the radar's range.

The UFOs had been circling the White House and Capitol Building for about two hours when at 2:00 AM the jets have arrived. However, just moments before the jets' arrival, the UFOs mysteriously disappeared. Incidentally, the radar stations were tracking another target north of Washington, but that too vanished. Then things got really bizarre when one of the controllers radioed Andrews to tell them they had a radar blimp right over their station. Then "the operators looked and saw a huge fiery-orange sphere hovering in the sky directly over their range station" (NICAP).

Back at Andrews, AFB that was tracking UFOs Staff Sgt. Charles Davenport reported seeing orangey-reddish light that would remain still and then abruptly change direction or altitude. At 4:30 in the morning, E.W. Chambers, a radio engineer, reported seeing five very large discs circling above in loose formation then went off into the sky (NICAP).

Our air force personnel tried to cover up at first claiming that Andrew wasn't tracking UFOs, jets were not sent to the area, or the equipment was faulty. Meanwhile, Blue Book was already busy producing many reports on the sightings. And the press printed fantastic accounts.

A week later July 19, 1952, the pilot and a flight attendant noticed unusual objects above their plane on their final approach into Washington. Within minutes, both National Airport and Andrews AFB were tracking once again several UFOs moving around at a hundred mph. One witness who saw them said they weren't like meteors because they didn't have tails.

Then on the 26th at 9:08 in the evening, the same controls picked up more UFOs. This time though, there were some top brass present as witnesses. The Pentagon's best UFO researchers: Major Dewey Fournet, Albert M. Chop for the USAF's press desk, and Lieutenant Holcomb, a radar expert. By 12:30 the next morning, four to five UFOs were being tracked. As before, aircrafts were ordered to scramble to the area (this time two F-94 Starfires), but also as before, the UFOs simply vanished from the radar screens just prior to the Starfires' appearance. But then reappeared again when the jets returned to base.

Then several people around Langley AFB called their base to report "weird lights that were rotating and giving off alternating colors" (NICAP). The tower at Langley also saw lights and called for a jet to intercept. The plane caught only a short visual and radar contact but witnessed nothing spectacular. At the end of the show, the radar sets were checked for any malfunctions, and none were found. Later when asked how he felt about his experience, Albert M. Chop, the USAF press man replied, "We all knew these objects represented something with which we could not cope" (nicap.org/wns2.htm). The Francis Ridge Report doesn't give much of an explanation but does state what it couldn't be. Based on its maneuvers, such as sudden stops, fast accelerations, or abruptly reversing or changing directions, the UFOs cannot be any forms of conventional aircrafts.

Whatever it was, though, it made quite a sensation and made headlines in many papers. The White House apparently took it seriously because President Harry Truman personally called Captain Ruppelt, the director of Blue Book, for an explanation of the events. Capt. Ruppelt's explanation was far from perfect, but probably still the more likely Earthly answer there is. Ruppelt reasoned that the radar sightings might have been caused by temperature inversion, where a warm moist air covers an area of cool dry air nearer the ground (Weatherquestions.com). These inversions can cause radar signals to bend and pick up something that is off by quite some distance. To find answers to the UFO question, the Pentagon held it most largely attended press conference since WWII held by Air Force Major General John Samford. Samford believed that meteors, stars, and other aerial phenomenon along with temperature inversions could explain the sightings. To lend support to this hypothesis are the crew of a B-26 bomber that was in the area at the time but saw nothing out of the ordinary. Blue Book also took a similar explanation that the visual sightings were meteors and the radar blimps were temperature inversion (Weatherquestions.com)

Many people strongly disagreed with the air force explanation for the sighting in July of 1952, from eyewitnesses, to air force personnel, to UFO experts. Capt. Ruppelt also disagreed with Samford's findings. According to Ruppelt (in Wikipedia blue book), there was barely a night in June and August without a temperature inversion, yet not a single UFO. Ruppelt also interviewed personnel at the control and radar towers, who witnessed

THE ORIGINS OF LIVING PLANETS

the events; not even one of them agreed with the air force's explanation. And when Michael Wertheimer interviewed them again 14 years later, they still disagreed. Furthermore, according to the International News Service (in Wikipedia), the U.S. Weather Bureau disagreed with the temperature inversion hypothesis. Even more interesting is that the crew of the radar tower claimed to have seen very large "fiery-orange sphere" (Wikipedia). However, when Ruppelt interviewed them a few days later, the crews claimed that they must have been mistaken and saw a bright star instead. But, Ruppelt checked with charts and found that there weren't any bright stars that night in the area. Ruppelt claimed to have good evidence of coercion from superior officers to say that they saw a star (Wikipedia blue book)

Many people were taking interest in the UFO phenomenon with the papers exaggerating the situation at best. The air force's response was to form a study (in addition to Blue Book) called the Robertson Panel to study the sightings including the ones in DC. The Robertson Panel concluded with just 12 hours of study that most of the sightings can be accounted for by natural phenomenon with the remainder being able to do so in the same way if more time was invested. The Robertson Panel recommended a public relations campaign to inform the public about UFOs, their explanations and assurance that they pose no threat to national security or personal wellbeing. The Robertson Panel did perceive an indirect threat, though. They feared that the intense interest in the subject would lead to a disruption from overwhelming military communications with frequent sightings or interest. The Robertson Panel believed that the public relations campaign focus on debunking the subject and reduce public interest (Black Vault).

The Bettelle Memorial Institute (a research organization founded in 1929) who did much more research into the investigation than the Robertson Panel, who only put in 12 hours, came up with a slightly different view. Although Bettelle found it highly unlikely that they came from higher intelligence, they worked for two years on the project and 3,200 reports. Bettelle found that from 18 to 35 cases were truly unidentifiable

Interpretations

So, what are the realistic explanations for these occurrences mentioned above? In many cases, different witnesses claimed to see similar objects from different positions, such as the Roswell, Mantell, the Chiles-Whitted encounter and other encounters, which rule out fictitious claiming or seeing things. The big question here is how could the witnesses see similar items all unexplainable in different places at times and didn't know of each other? Add to this is that sightings often report similar objects, such as cylindrical, disk or cigar shaped flashes of lights, but never with, as far as I know, wings. I don't know the answer for these similarities, but what I do know is we shouldn't assume the ETI hypothesis either. Many sightings can be explained with aerial objects, such as meteoroids, Venus, stars, temperature inversion, balloon, or even birds with reflective bottoms. But, how about the really tough cases that were found unknown that have been estimated at around ten percent of cases, especially those with reliable eyewitnesses?

The similarities between sighting are difficult to explain with natural phenomenon but not impossible. The similarities of sightings often reported include flashes of light, disks without wings, moving very fast or making quick sudden turns that conventional aircrafts can't perform because of the G-force, pilots would pass out. According to my own counts, there were three witnesses of bright lights, four elongated or cigar like discs, several supports with radar and numerous supporting documentations with eyewitness who all reported a similar sighing. The radar evidence can be easily explained by temperature inversions, but how about all these corroborating witnesses who all reported elongated discs or flashed of light?

The answer to the eyewitness problem is that eyewitnesses are notoriously unreliable. Eyewitness testimonies have been the foundation of jurors decisions in the court system for decades but have been recently found to be unreliable through science. Jurors really put a lot of faith into credible eyewitness testimony, which decides many cases. However, just because jurors find eyewitnesses reliable, doesn't necessarily mean that they are. The human mind does not record events like a tape recorder, or a computer would. Instead, human memory is more pieced together afterwards and is subject to influences from the present and past.

Loftus, E & Zanni, G. (1975) did some of the first, as I know, experiments on eyewitness testimony. They have shown how a simple choice of wording in a question can distort memory; in this experiment, the word choices were between the words *the* and *a*. After watching an accident video, the participants were either asked, "did you see a broken headlight", or...the broken headlight. The participants had three choices; yes, no, or I don't know. Also important is that some of these questions were pertinent to what happened in the filmed accident and others were not. Participant with the *a,* or general article were twice as likely to answer "I don't know" in both the pertinent and non-pertinent questions, as opposed to the specific article *the*. Also, the specific article was twice as likely to elicit a yes answer when the question did not pertain to the video, such as did you see a flat tire, or such. The authors also did another experiment in the same way except for the video itself was of another accident. This was done in order to replicate the study, and to confirmed that the phenomenon is general for all contents and the psychological make-up of people. The results were duplicated with a different videos and participants. What this demonstrates is that even the difference between two words can influence what people remember from an accident. And very important, these subtle changes with big consequences can apply to UFO sightings as well.

 A more recent study supports Loftus, et al (1975). In 2004, Mudd & Govern did similar research into eyewitness memories, but this time Mudd, K., & Govern, J.M. (2004) used a video about the police pulling over a car, but the car sped away and escaped. Before the video, the participants, who were tested individually, were introduced to a confederate, but the participant thought she was a participant. After the video, the confederate in the "misinformation" group initiated a conversation that led to her stating that she thinks the case was a "stolen car," and in the control group she started a conversation about the weather and that's it (Mudd, et al). In the control group, one of the participants answered stolen car while six answered to this question in the misinformation group. While in the "I don't know" answers, there were 12 in the control and only eight in the misinformation group. What this research indicates, as did Loftus et al (2004), is that memory for incidences can easily be influenced by present events, language, or suggestions.

Vertigo is one explanation SIGN has given for some of the sightings by pilots, such as the Mantell sighting (nicap.org/sign). The American Heritage Dictionary defines vertigo as dizziness and the false perception that either you or your environment is moving. Consider this and the fact that many pilots reported making sharp turns, banks, and altitude adjustments to view the UFOs, it is not a far stretch to imagine that even an experienced pilot could be affected by vertigo and this influenced them to see an unusual object, such as a balloon as much more extraordinary. As for when both the pilot and co-pilot reported a similar setting, SIGN reported in 1949 that because both pilots received the same exposure to altitude changes, G forces and such, it explains why they both saw similar sights.

Along with aerial objects, stars, and Venus and temperature inversions, the way memory is formed and later recalled is all variables, or contributing factors, to explain the sightings. I'm not ruling out ETI, though, because some sightings are convincing and very difficult to explain even with the answers I just gave. However, isn't temperature inversion, or distorted memory just as probable explanation than ETI, if not more? After all, both may be unlikely in some of the cases, but even the unlikely occurs some of the time and there must be explanations.

Another explanation for these sighting is mass hysteria and the desire to believe. These sightings have been accused of threatening national security. Even President Truman had been notified in one case. Another case of mass hysteria is the Red Scare, where many, including those in Washington, believed the UFOs were Russian aircrafts sent here to illicit mass hysteria in an attempt to destabilize our society and government. I have a couple of problems with this explanation. First, it sounds very paranoid. In fact, the whole Red Scare (largely perpetrated by Senator McCarthy) was largely a result of fear and paranoia. Senator McCarthy tried without success to expose Russian Communist and accomplishers in trying to destroy our way of life and government. There were a quite a few others involved as well. The fact that they have never come up with even one piece of evidence I find as evidence of a wild goose trace and paranoia.

The Continued Search for ETI

Neither science nor the government has given up on exploring extraterrestrials. I suspect that many want to believe. After all, wouldn't it be exciting to be visited by alien life forms far different than ourselves and from billions of miles away? They just go about it in different ways. For instance, Pioneer 10, launched March 2, 1972, was equipped with a plaque with a short message and guidance to our sun and planet (quest.nasa). Pioneer II was launch shortly after with another close encounter plaque and was the first to Saturn. Pioneer 10 was the first to leave the solar system on June 13, 1983, when it passed our outermost planet at seven billion miles from the sun (quest.nasa).

Another way we are searching for extrasolar life is by monitoring the cosmos with powerful radio telescopes. A private run ETI research institute is SETI, or Search for Extraterrestrial Intelligence. Once cut off from government funding due to thinking that it was a search for "little green men," the organization is now thriving through private donations and grants, such as the National Science Foundation (Washingtonpost.com May 30, 2005), through the NASA Ames Research Center and more (seti.org). In addition, SETI now publishes in scholarly journals. Although they haven't found any radio signals emanating from extrasolar systems, it hasn't caused them to stop. In the first stages of development is a very powerful radio telescope that may show great promise once it starts looking. Named after its most generous sponsor, Paul Allen, the telescope is an array of 350 radio telescopes working in unison to act like one giant telescope that increases their sensitivity by a whole bunch (Wash. Post, same date). The telescope analyzed one star at a time (chosen for life supporting possibilities) and has scanned about 1,000 stars, as May 30, 2005.

A NASA project that is looking for small Earth-like planets is the Kepler mission (kepler.nasa.gov). Kepler is a space probe with a telescope designed to detect tiny light changes when viewed from afar, which is called the "transit method" of detecting planets (Kepler). When a planet passes in front of its star as view from a distance, it blocks out very small amount of light that Kepler is scanning for. With this light change measurement, Kepler can calculate the mass, density.

Rebuttal

Maybe the most convincing case of all is the Captain Mentell case because of cooperating, lack of reasonable explanations and agreement of witnesses on what happened. However, I see the balloon hypothesis along with lack of oxygen on Captain Mentall's part along with some self-hypnosis partially due to oxygen deprivation. Although balloons don't move at high rate of speed as reported by Mentell and other witnesses, hypnosis could theoretically give the appearance that sightings are real. As far as corroborating witnesses, this is difficult to argue against because witnesses in different locations witnessed similar events, bright lights moving fast. Maybe these bright lights were reflections from the balloons, or maybe it was another phenomenon to what Captain Mentell witnessed, the balloons. Whatever the explanation, or explanations are they certainly more plausible than ETI from 100,000 light years away. Remember the speed of light is the speed limit in the universe. This means that even if traveling at the speed of light (which is unlikely due to $E=MC2$), they would take a hundred thousand years to reach us

THE ORIGINS OF LIVING PLANETS

Yet another difficult case to find answers to is the Roswell case. Mr. Brazel found some metal pieces on a homestead where he worked as a foreman and took them to the U.S. Air Force for examination who have determined it to be metal from a balloon. To me that settles the case. However, the media evacuated the case and blew it out of propulsions. This is a good example on how people can make things appear to be from out of this world. People seem to want to believe that UFOs are from space aliens and will unconsciously distort the facts to make it appear so.

CHAPTER V

ALIEN ABDUCTIONS

Case I

ABDUCTIONS ARE FASCINATING PHENOMENON that can probably all be explained through scientific explanations or as products of the imagination or fictious. One of the first published, captivating and explainable (with present day knowledge) abductions was from Betty and Barney Hill in September of 1961. Barney Hill was a Black U.S. Postal worker who was married to Betty, who was a white, Master's degree, social worker, and interracial marriages were rare and shunned in the early 60s. At first it was thought that the stress of such a racially mixed stigmatized marriage caused them to imagine things, but this hypothesis was quickly abandoned once they interviewed them and found both of seemingly sound mind.

It started around 10:00 in the evening in New Hampshire while driving home when Barney and his wife Betty noticed what first appeared to be a "bright star" (ufo casebook, & Wikipedia). They made eye contact with it for several minutes while thinking it was an airplane. Later, after they witnessed it fly in strange ways, Barney stopped the car to have a good look at the light through his binoculars. What he saw were lights in many colors, rows of windows, the craft itself was rather flat, and there was no mention of wings. The object seemed to be moving towards them, and when the craft got close, he claimed he could see occupants through its windows. Frightened, Barney quickly ran back to his car where his wife was waiting, and the couple quickly raced away. Shortly after driving a while, they heard an unidentified beeping sound twice, and then suddenly realized that they were 25 miles further down the road while only a couple minutes prior (ufo casebook, & Wikipedia). Later the next day, Betty reported the incident to the U.S. Air Force and spoke with Major Paul W. Henderson, who notified her that the UFO was also spotted on their radar.

Now this is where the story starts to get suspicious. Betty started to have disturbing dreams of Barney and herself on an odd aircraft. Then a couple writers got wind of the UFO sighting and did some interviewing. They discovered gaps in their encounter of two hours, which were completely unaccounted for. Then to seek answers, treatment for their trauma, and to fill in for the loss time, they sought out hypnosis. They were referred to a well-respected Boston psychologist and neurologist Dr. Benjamin Simon. Dr. Simon subjected the Hills to six months of hypnosis and then determined that they were abducted and gave details of the event. According to their "recovered memories," they remember being on the road when the car stalled, the craft landed on the road in front of them and then they were taken onboard. The aliens, as they reported through hypnosis, were around five feet tall, slanted "cat-like eye," grayish skin and large "pear shaped" heads (ufo casebook, & Wikipedia). The aliens then, Barney and Betty claimed, took them onboard their ship and did medical tests, including what they claimed was a pregnancy test on Betty and a semen sample from Barney. After the medical tests, the aliens escorted them back to their car. Then the aliens left them with a big glowing orange light as their craft flew off.

Explanation

Hypnosis often doesn't only help people remember repressed events, they can often instill false memories (FMS Foundation). False memories or false memory syndrome (FMS) are memories for the past of something that never occurred, and the person afflicted often believes these false memories very strongly despite evidence to the contrary. This is what Psychologist Robert Baker stated about hypnosis to recover lost memories, "An overwhelming body of research indicates that hypnosis does not increase accurate memory but does increase the person's willingness to report previously uncertain memories with strong conviction" (in FMS Foundation).

More Cases

There have been many cases of so-called "recovered memories" in the courts back in the 1980s and 90s (Perry,

D., Gold, A.D. 1995; Kaplan, R., & Manicavaagar, V. 2001). They were mainly based on "recovered memories" as an adult remembering back to her childhood of her father molesting them and most were obtained under hypnosis. One tragic case was that of Paul R. Ingram in 1988. Mr. Ingram was accused and sentence to prison based on charges of bizarre satanic sexual rituals. Mr. Ingram allegedly led many other cult members in satanic ceremonies that sexually invaded his daughters. What's more terrifying is that the Ingram case wasn't that rare, for the courts were using memories obtained under hypnosis as testimonies in criminal cases of incest. Even more is that many of these cases, the accused was convicted solely on these testimonies. I bring up this case to show how "recovered memories" are not recovered but are false memory syndrome and this FMS can be implemented in alien obductions. Some people are more vulnerable to hypnosis and creating their own FMS though self-hypnosis.

More Explanations

The psychologists weren't malevolent when they instilled false memories into their clients. They were following the scientific community's general concepts about "repressed memories" and recovering them. According to Freud, severe trauma is so difficult to deal with that the child unconsciously represses the memory deep into the unconscious portion of her psyche. He first believed that repressed sexual abuse could explain most of his hysteria patients, but later revised his theory into the

Oedipal complex (Kaplan et al, 2001). In the 80s and 90s, psychologists believed that repressed memories were at the heart of many emotional problems. What's more, professionals believed that these repressed memories could be retrieved through "Recovered Memory Therapy" or RMT," which usually involved some form of hypnosis (Kaplan, R., & Manicavasagar, V. 2001).

The same kind of FMS can occur with alien abductions. Just like sexual abuse hypnosis, similar effects can create false memories of a variety of things, including alien abductions. I don't see how hypnosis can produce false memories of incest and satanic cult abuse and not false memories of being abducted. Both can create FMS. Also, about 15 percent of the population is extremely acceptable to hypnosis, and I believe that many of the abductions are from this group. Therefore, I see false memory syndrome as the most probable answer to Barney and Betty's case and to many other alien abductions.

Another Case

Another out of this world encounter that also involved hypnosis started around 6:30 in the evening on January 26, 1967, in South Ashburnham, Massachusetts. Betty Andreasson was in the kitchen while her seven children and both parents were in the living room when the lights in the house blinked on and off, and then a "reddish light" beamed through the kitchen window. As Betty rushed to comfort her frightened children, her father ran into the kitchen to discover the source of light.

What he claimed to see will knock the socks off most

THE ORIGINS OF LIVING PLANETS

open-minded people. He stated to have seen a saucer-like ship and five odd beings hopping like bunnies toward his house (UFO Case Book). These "aliens" had pear-shaped heads, like the Hill encounter, and hopped right through their wood door. Then the beings put the whole family into a state of "suspended animation" (UFO Case Book). One of the creatures approached Betty and made telepathic communications with her. Another thing that this encounter with other, besides pear-shaped heads, was the height that they were rather short around, five feet, that the tallest one seemed to be the leader and the shape of the craft was disc-shaped.

In 1977, when the case was first seriously investigated, a team of experts was assigned to examine the case who subjected her to many test including "character analysis, polygraph examinations, psychiatric reviews, and 14 sessions of regressive hypnosis" (ufo.about). The hypnosis supposedly revealed more about the abduction that she repressed.

Betty claims the whole family was in a state of suspended animation that is kind of hypnotic-like in character. The Illustrated Encyclopedia Dictionary defines suspended animation as "a dormant condition resembling death, induced by suspension of the vital functions." The same dictionary defines hypnosis as "an artificially induced sleeplike condition in which an individual is extremely responsive to suggestions…" Now it seems that suspended animation is awfully chose to hypnosis; in both cases, the person is not fully conscious. So, maybe what Betty and her father were both experiencing was hypnotic trances brought on by some unusual, but Earthly lights, which made her suggestible and created false memories. Therefore, the only things real in Betty's memories is the lights that could have come from human activity of some kind, and this triggered a trans-like state that facilitated false memories of events that never even remotely took place.

Physical Evidence

You may be asking, though, if these abductions are real, then where is the conclusive physical evidence? Surely aliens couldn't be abducting our citizens below our noses without any evidence to prove they been here. I'm only aware of one case of alien abduction with any physical evidence at all, and this is the Peter Khoury case in Australia. This is a very odd case from the start, but at the same time very festinating. Mr. Khoury had a long history of seeing aliens, starting in Lebanon, his homeland, in 1971 when he was only seven years of age (Chalker, B., 2005). He claimed to have been going up on his neighbor's roof when he saw seven other children "frozen like statues in front of him, while a silent egg-shaped craft hovered above" (Chalker, B 2005). Then without any memory of getting there, all the children found themselves on the ground floor.

Then in 1988 on the 12th of July, Mr. Kroury had another experience. He claims that while resting in bed, he was aroused by several human-like creatures. One with golden-yellow skin and "large black eyes" unsuited a long needle into the side of his head at which time he blacked out (Chalker, B 2005). After awaking, he felt anxious and rushed into the next room. There he saw his family all in a trans-like state, similar to other claims of encounters. He was able to arouse them though, and when he did, they all thought that only a few minutes have passed, but in reality, nearly two hours had past. Chalker never mentions, though, whether a whole was found where the needle supposes to have entered, so I assume that there weren't.

His next experience is the strangest of all and led to one of the only physical evidence of aliens that I'm aware of. It happened on July 23, 1992, around 7:00 in the morning after returning from work. He felt sick and laid down on his bed for a while. A few moments later, he was startled awake and noticed two naked strange looking women kneeling at the end of his bed (Chalker, B., 2005). One was Nordic looking while the other looked Asian, but both were different looking. The Nordic woman had an extremely elongated face, a sharply pointed chin, blue eyes that were two to three times larger than one would expect and "very fine wispy blonde hair" that seemed puffed up" (Chalker, B 2005). The Asian appearing female had darker skin and just about completely black eyes. The Nordic woman, who was over six feet tall, seemed to be in charge. She tried to push Mr. Khoury's face into her breast, but he resisted. After her third try at this, Mr. Khoury took a bite out of the Nordic woman's nipple and swallowed it. She did not show any sign of pain or bleeding, but she did seem quite shocked by this. Mr. Khoury then had a coughing spasm, and when he looked back up, they were gone, just like that, both just vanished.

Then Mr. Khoury's coughing drove him to the

bathroom to get some water. There he discovered maybe the first physical evidence of an alien close encounter: two hairs of the Nordic looking woman, which he placed in a plastic bag and sealed. He knew it was the Nordic woman's hair because it was unlike Vivian his wife's hair. Like other abductees, Mr. Khoury had many more strange and unexplainable events. Some were sightings while other times he just felt an energy source. If it was my experience, though, I would have this hair examined ASAP, but for some reason, Mr. Khoury kept the hair and did not seek out answers from it and even kept it a secret from his wife for two weeks.

In 1996, Mr. Khoury, like other abductees, was hypnotized. Under the care of Harvard psychiatrist John Mack, Mr. Khoury gained more memories of some of these experiences. But still Mr. Khoury kept the hair sample a secret from the world. Maybe he was afraid of how others might react that kept him from revealing the full story, but in the long run it doesn't matter because hair samples keep.

It wasn't until 1998 that Mr. Khoury made public his discovered hair sample. The world's first DNA profile a polymerase Chain Reaction, or PCR test was performed on these two hair samples (Chalker, B., 2005). According to Mr. Chalker, the test showed Khoury's bedroom visitors to be "chose to human," but one of the rarest lineages in the world, a "Chinese Mongoloid type." Mr. Chalker (2005) says, "close to human," but he was never really clear. Furthermore, I can only imagine that the creatures were either human or they weren't, not some hybrid or close lineage to us. Therefore, the most reasonable explanation was that this Nordic looking woman was of very rare linage, but still fully human.

In addition, further DNA anomalies were found to make her even more unique. According to Chalker, the first analysis was done on the hair shaft, where they found her to resemble a Chinese Mongoloid, but they performed another analysis on the hair root and found an also rare "Basque/Gaelic type DNA." The answer to this came in 2000 when they determined it could have come from some kind of hair transplant (Chalker, B). This makes this woman even more rare than before, a Nordic woman with a hair transplant, but I still find her identity to be human as the best plausible explanation. What's more, if this evidence were conclusive, then it would be all over the media, but it's not. I suppose it could be that many people may be resistant to accepting space aliens, but I see the more plausible explanation is that the scientific community haven't taken the Mr. Khoury DNA sample that seriously because the most probable explanation is that she was a very rare linage of human.

An interesting notable observation is that most of Peter Khoury's experiences were in his bedroom. Maybe in his bed, he is more likely to experience a trans-like state that can be like a hypnotic state where one is very suggestible and vulnerable to forming false memories. Chalker B. (2005) stated that the 1988 experience was unlikely "sleep paralysis or hypnagogic imagery" because his encounter started as soon as he laid down. However, it is not unheard of for certain people to quickly fall into hypnosis and there's the possibility that Mr. Khoury lost some track of time after laying down. This may sound like a farfetched explanation, that he can rapidly go into self-hypnosis by simply laying down, but isn't this more likely than ETI?

The big question is, what does all this data we have including eyewitnesses and DNA lead us. Well, my answer is very similar to how the universe and living planets develop, which is through chance. That is, even the very unlikely is probable with enough chances. Planets develop the right size and distance from their sun by chance, life starts by chance, and given the millions of people in the world, it is only logical that some of them will develop aberrant features, like predisposed to hypnosis and its effect, and have unusually strange things happen to them, such as meeting rare women. All of these may be extraordinary Earthlike explanations, but with enough chances all most anything is possible.

In case one, the Barney and Betty Hill abduction, I mostly explained it through FMS induced through hypnosis. I suspect that both Barney and Betty were one of the 15 percent who are highly susceptible to hypnosis, or other trans-like states. The later part, memories recovered under hypnosis by Dr. Simon, can entirely be explained through FMS. I also believe that their earlier experiences while driving can also be explained mostly with FMS and trans-like experience. Self-hypnosis is nothing new, and there's even at least a couple of self-hypnosis videos on YouTube (Kayla, Garnet, & Rose). These videos ask the person to stare into a circle spinning video while listening to a voice while wearing earphones. I suspect that it is possible for a person or even two or more people to become hypnotized by staring at a bright light

THE ORIGINS OF LIVING PLANETS

in the sky. As evidence, they did report watching it for several minutes before they seen anything odd. Therefore, I determine that the most likely case scenario is that they did see some odd light, which after staring at it for several minutes, Barney and maybe Betty too were hypnotized, and this created false memories of abduction.

In other cases as well, I suspect highly susceptible individuals were hypnotized and false memories were formed. Take the Betty Andreasson case again for example. Betty and her family reported first witnessing a bright light shine through their kitchen window, which could have put them all in a trans. This multi-hypnosis is not that odd. We all heard of hypnotists who put multiple volunteers into hypnosis and have them make fools out of themselves in front of an audience. If mass-hypnosis like this is possible by a hypnotist, then accidental self-hypnosis on multiple persons ought to be possible in rare cases.

Now let's examine the Khoury case. The possibility of this case occurring is probably one out of millions, but still happened. The first odd occurrence was Mr. Khoury's high susceptibility towards hypnosis, with most encounters happening in his bedroom where he could more relax and experience hypnosis. Then there was the unlikelihood of meeting two very rare human. All this may be very unlikely, but it is the best explanation there is, and given the millions of people in the world, there's plenty of room for the very unusual. Moreover, even though this explanation seems farfetched, it is certainly more plausible than the ETI hypothesis.

CHAPTER VI

PLANETS' LIFE CYCLE

The final question of living planets may be their destinies. There are probably many demises that can befall a living planet. Many species may linger on for thousands or millions of generations without advancing out of primitive stages. Other species could keep advancing till they are thousands of years advanced in technology and knowledge than we are. Or there could be some natural catastrophe like the giant meteorite that struck the Yucatan Peninsula and wiped out most of the dinosaurs and other life forms. Incidentally, some of the dinosaurs survived and evolved into birds, there are a lot of similarities between birds and smaller dinosaurs, such as when walking they both walk on two legs with talons on the ground. I also see another possibility that species simply destroy themselves as they advance in technology and population, as we seem to be doing.

There is evidence for this last possibility right here on Earth. For instance, we have weapons of mass destruction that can destroy all living life many times over. Our population is expanding at an exponential rate. And then there are the environmental issues that might eventually sneak up on us. If humans are only moderately violent relative to other species in the universe, then there are many that are more violent than us who might use their own weapons of mass destruction to wipeout their planet. And this may be how some living planets may destroy themselves.

A kind of self-destructive demise that may befall advanced life forms are the destruction of their own life-supporting atmosphere. We may witness this on Earth with the destruction of global warming and may eventually cause catastrophe if we don't change our ways drastically and soon. The runaway greenhouse effect may eventually occur where the slow but steady increases of average global temperatures takes on a momentum of itself without us putting more greenhouse gases into the atmosphere. According to nasa.gov, when the global average sea temperature rises above 80 degrees F, a critical point is reached where more heat is absorbed than what is radiated back to space. In other words, the "hotter the surface temperature gets, the faster it warms up." This runaway effect will occur when enough water vapor enters the atmosphere through evaporation that heats up the atmosphere. Water vapor allows solar energy to easily pass though, but traps much of the energy. Since evaporation increase with temperature and more atmospheric water vapor traps more heat, it creates a runaway effect. Another potential contributor to runaway global warming is if the polar caps melt, less heat is reflected back into space that will warm the Earth further. According to NSIDC, the Antarctic ice shelves have been recreating since the 1970s.

I'm convinced that some extraterrestrials have already met one of these fates, while some species may have worked around it. Maybe the less aggressive species are less likely to blow themselves up, or species that have enough foresight to avoid thermal runaway and advance indefinitely till they are much more advanced than us. However, for them to do so, they must learn to live in harmony with their environment and not be so aggressive to destroy themselves.

Let's not be another statistic within the universe and become one of those more aggressive beings that destroys themselves through the many options at our disposure: climate change, species extinction, habitat destruction, over population or weapons of mass destruction. Rather, lets become one of the more intelligent and foreseeing life forms that prevents our own self destruction. We can do this by controlling our reproduction rates and limiting our CO_2s output and encroach less on habitats.

PART II

THE PLANET'S WELLBEING

CHAPTER VII

CLIMATE CHANGE

GLOBAL WARMING IS ABOUT changes in global average temperature. It is not about local fluctuating's temperature. Local temperature fluctuates according to many variables. However, global average temperatures have been steadily rising for the past century or so—ever since humans have been dumping greenhouse gases into the atmosphere. There is overwhelming proof of this rise, such as temperature readings, polar ice cap melting, more frequent and severe storms, realized predictions, and rising carbon dioxide. I will first start with the evidence for global warming. Then talk about the reasons behind global warming. Then, I will talk about the solutions. Finally, I will talk about other threats to the planets and solutions for them.

Global warming is caused largely by rising carbon dioxide (CO_2) levels. CO_2 rises are caused by many things and are mostly due to human activity. One of the biggest human-made rises in CO_2 are automobiles and factories including factory farms. According to U.S. Publisher Ward's, there were 1.015 billion automobiles in the world in 2010. What's more according to Iowa State University, a single vehicle will produce at least 30,000 parts per million (ppm) of CO_2 in its lifetime. Methane another greenhouse gas comes from animals especially from factory farms. Because animals are kept near one another in factory farms, their output of methane are concentrated and higher. According to Moss, A.R., Jouany, J.P., & Newbold, J. (1999), the EU alone has 10.2 million tons per year of methane. Factory farms are an attempt to feed an ever-increasing human population, which is another problem I will touch on later. However, animal farms are a poor way of feeding our ever-expanding population.

These farms are cruel and unethical because of the way animals are raised in very close captivity, laying in their own wastes. For instance, chickens, who PETA claims are the most tortured animal on the planet, must have their beaks removed to prevent them from pecking each other to death. This is totally against their nature and is a direct result of being so close to one another. There are numerous examples of cruelty to animals within these factory farms. Also, farms must use antibiotics because of close proximity. Furthermore, these farms use growth hormones to increase the animals' weight and meat outputs. They also use antibiotics to decrease diseases that are made more likely because of close proximity. This is unhealthy because over use of antibiotics increases immunity to them and these chemicals are in the meat we eat. The question is, are these factory farms necessary to feed our ever-growing population? The answer is no, if we take certain steps. The first step is to eat less meat. According to Kristi Wempen at the Mayo Clinic,

Americans are getting twice the amount of protein they need. Also, according to the Mayo Clinic, only 10 to 35 percent of your diet should be of protein. Plus, there are other sources of protein, such as quinoa, beans, soy, and nuts. Then there is the vegetarian diet that millions live off. This diet is totally void of meats, but totally sufficient for supplying the necessary nutrients to sustain good health. I'm not suggesting everyone become a vegetarian, but only that we don't need to eat nearly as much meet as we are consuming.

Other contributors to global warming are chlorofluorocarbon (CFCs) and nitrous oxide (Pearce, D). CFCs are organic compounds composed of carbon, fluorine, and chlorine. Whenever CFCs contain hydrogen in addition to one or more chlorines, they are called hydrochlorofluorocarbons, or HCFCs, (Britannica). Although CFCs are small in the atmosphere measured in parts per trillion, they are a major contributor to global warming (Enviopedia). This is because CFCs are very good at trapping heat in the atmosphere. On a good point, because CFCs effect as a green house gas and on the ozone layer, there use have significantly been reduced.

Whatever the causes of global warming, it is irreversible, and at the rate we are going, it may be too late to stop it soon (Pearce, D. 1991). However, we can take steps now that slow or stop the advances of global warming that I will talk about later.

Another big culprit in the case of global warming is factory farms. Not only are factory farms amoral and cruel to animals, but they are also a major contributor to global warming. For one thing, factory farms take up space, which strips the forests of trees that are CO2 eliminators (Compassion in World Farming). They also produce vast amounts of manual, which are also CO2 producers. They also produce a large amount of fertilizer and pesticides by cramming so many animals into a tiny space. Also, antibiotic must be used because animals are kept together in tight places that means they can get sick from one another. By cramming many animals into a small space that is not only cruel and unjust but also causing harm to our environment.

The Arctic ice caps are slinking in area exponentially (NASA). The ice caps wax and wane with accordance to the seasons, expanding in the winter months and shrinking in the summer months. However, the ice caps in the winter are not reaching the expansion in the winter and shrinking more in the summer compared to previous years, and this is intensifying per year beginning when measurement started in 1970s and have been shrinking every year since. Parkinson, & DiGiolamo (2016) have measured that since the 1970s, the Artic sea ice have has shrunk on an average of 21,000 square miles, or 54,000 square kilometers with each passing year. This is equal to losing a chunk of sea the size of Maryland and New Jersey combined each passing year.

However, the thickness of the ice sheets is at question, but with a new satellite ICWARat-2 launched by NASA this year, 2018, more data will be available on the thickness of the ice sheets. This will add more data to our estimate on global warming. However, with the data we already have it all points to the fact that the planet is warming.

Furthermore, according to NASA's National Snow &

Ice Data Center, the Arctic ice sheets are at a 40-year low. Clearly there is over whelming evidence from the Artic ice sheets alone to indicate that the planet is warming. The Arctic ice sheets have been steadily, year by year, receding. What's more, that decrease has been occurring along with the rise in CO_2 and other greenhouse gases increase. Therefore, considering that the Artic ice sheets have been receding for the last 40 some years and there's no other explanation other than rise of greenhouse gases, and the greenhouse gases are being put there by human activities, it seems prudent that this is caused by human activities.

For skeptics, I suppose the correlational evidence between rises in greenhouse gases and global warming and shrinking ice sheets are just that correlations and don't necessarily equate to cause and effect. It is a scientific fact that correlation don't necessarily equate to causation, for there could be another variable involved. For instance, there is a positive correlation between children's reading abilities and their shoe sizes. They both rise together. However, no sane person would suggest that larger shoe sizes are contributing to higher reading abilities. This is an example of a third variable, that is children are growing and with growth, there is improvement in reading and larger shoe sizes. The third variable then influencing both is growth.

First, what kind of third variable could there possibly be that contributing to global warming and greenhouse gases? I can't imagine one. Plus, the greenhouse gases have been only increasing exponentially within the industrial and postindustrial areas. Furthermore, this rise has been contributing to many predictions, such as the retreating of the Arctic ice sheets and more frequent and severe weather conditions, such as hurricane, tornadoes, and draughts. Moreover, we know that greenhouse gases, especially CO2, raises global temperatures. Take Venus, for example, Venus is on average 800 degrees plus Fahrenheit and Venus has a concentration of CO2 many times greater than that of Earth. Greenhouse gases like CO2 and CFCs trap the heat in the atmosphere that rises of the Earth global temperature. The sun heats the Earth and the Earth then radiates heat back up into the atmosphere out into space, but greenhouse gases prevents the heat from reaching space and instead remains trapped in the atmosphere further warming the Earth. There is a correlation between rising greenhouse gases and heat radiation escaping the atmosphere. The greenhouse effect from water vapor is filtered out, showing the contributions of other greenhouse gases (skepticalscience.com)

Furthermore, CFCs are greenhouse gases and destroys the ozone layer. The ozone is about 6.2 miles above Earth and protects us from the ultraviolent (UV) radiation from the sun. UV radiation is damaging to life on Earth. Excessive exposure to UV can cause skin cancer, sunburns, cataracts, suppression of the immune system, and genetic damage (Wikipedia). Moreover, the ozone is our main protection from UV radiation responsible for about 97 percent of absorption. Ozone works like this on UV radiation: When UV radiation strikes CFC molecules a carbon-chlorine bond breaks, producing a chlorine (CI) atom. The CI atom then reacts with ozone (O3) molecules breaking it apart and destroys ozone molecules (Wikipedia).

THE ORIGINS OF LIVING PLANETS

Methane is another big contributor to global warming. Two thirds of atmospheric methane come from farming: two thirds from intestinal fermentation and one third from livestock manure. The European Union dumped about 10.2 tons of methane into the atmosphere in 1990 (Moss, A.R., Peierce, J., & Newbold, J. (2000). As previously mentioned, factory farms are a major participant in methane because of the concentration of animals in small spaces, contribute to global warming. Also, according to Moss, A.R., et al. (2000) atmospheric methane is increasing by 30 to 40 tons per year. This is clearly the wrong direction; we should be decreasing methane production not increasing it. I blame this on the rising population and our increase in factory farms to maintain our ever-expanding population.

The Solutions

All is not hopeless though, for if we act now, we can halt the progress of global warming before it's too late. However, we must act immediately because global warming effects have already begun, loss of Artic ice sheets, more severe and frequent hurricanes, tornadoes, draughts, and flooding. If we don't act now, these atmospheric effects will greatly impair our earth, destroy wildlife, create mass starvation, flood coastal cities and countries, and threaten national security. Whole countries could go under water and be destroyed, such as Denmark and the Netherlands, and New York City could be lost. However, if we act right away, we can slow or even halt the process. We can never reverse global warming because we cannot distract greenhouse gases from the atmosphere. Therefore, we must start now before the damage becomes too great. If we want to save our unique planet from irreversible damage, we must act immediately. There is no room for waiting or contemplation.

There is no one solution for this problem, but there are multiple techniques at our disposal. One is solar power. We can put up solar panels on rooftops of buildings. When on rooftops the panels are often not visible and don't create an eye sore. We can also put them up in vacant fields and on light poles. Anywhere where there's unrestrictive sun rays without creating an eye sore. Another solution is wind power. As with solar panels, we could install them on rooftops, fields, etc. Another green energy is geothermal energy. This is heat from just under the Earth's crust deep under the surface sometimes all the way down to near the core. We can tap into geothermal energy for heat or producing electricity by tapping in steam and or hot water beneath the Earth's service in underground reservoirs. In the western U.S. especially underground reservoirs are common (google). Tapping into this energy works with pipes buried at least four feet underground. Then a liquid is pumped through the pipes to absorb this heat and bring it up to the service and then uses it to heat the atmosphere within a home, or it can be used to cool the house during the summer in a reverse method. Not only is geothermal energy green energy, but it is also efficient and cheap, (alliantenergykids.com).

This is tapping into the energy of the Earth, which originated from Earth's formation in the early years of the planet's formation (alliantenergykids.com). Even though

THE ORIGINS OF LIVING PLANETS

it's pricey to install, about 20K, it can save money in the long run with upward savings of 2k a year. Hence, after 10 years, you can start saving and get a profit out of your system.

A carbon tax is also necessary to taper off from carbon fuels. A carbon tax is considered by many to be the only way to get industries to reduce their carbon footprint and reduce their usage of oil, gas and especially coal. A carbon tax works like this: Anyone burning carbon fuels must pay a tax starting where the fuel is extracted from the Earth or imported from foreign countries to the burning of that fuel. Users are free to pass along the cost of the tax. Placing a tax on carbon fuels creates an incentive to burn less fuels and depend more on green energy mentioned above, (carbontax.org)

CHAPTER VIII

SPECIES ENDANGERMENT

THERE ARE A QUITE a few reasons why a species become endangered: loss of habitat, loss of genetic variation, pollution, introduction of an alien species into a new environment, poaching, destruction of natural environment or illegal hunting. An alien species being introduced to a new environment can take over the new environment because it doesn't have any predators to keep it in check. Then it over runs the new environment and chokes out the original inhibitors. Loss of genetic variation occurs when the species pool of members is too low to produce offspring with any diversity. Loss of biodiversity can be caused by habitat destruction, deforestation, overpopulation, pollution and global warming. Loss of biodiversity can mean that the species can't evolve in changing environments and can lead to endangerment or extinction.

Species are going endangered or extinct at an alarming rate. Possibly the most heard about and tragic species is the elephants. Poaching of elephant for the sole purpose of the tusks. Poachers kill these stunning animals, cut off their tusks and leave the meat to rot. This is an immoral act because not only do we waste their meat that could be used to feed hungry people, but we are destroying them to near extinction to make a profit. What's more, according to the Guardian, there are approximately 20,000 African elephants killed per year. In 1800, there were about 26 million African elephants. Now it is estimated to be only 50,000 African elephants remaining (the Guardian). Now if there 50,000 remaining, and we kill 20,000 per year, it means that the elephant will go extinct in two and a half years.

Another severely endangered species is the gorillas. Gorillas are giant apes and are seriously endangered for much of the same reasons as elephants are: poachers and infringement on habitat. Poachers kill these beautiful animals for many reasons, such as their meat, body parts, trophies or capture them for pets (Volcanoes National Park). In certain cultures, the body parts or eating their meat are considered healing medicine (volcanoeparkrwanda.org). As far as pets, wild animals seldom make for good pets. What's more, these "pets" can be dangerous to their owners and others. This is not only inhumane but foolish and dangerous. According to the World Wildlife Fund (WWF), killing for meat is the gorilla's greatest threat, and this is because of some folk beliefs about their meat being medicine without any scientific fact, but instead is based on ignorance. Also, according to the WWF gorillas share 93 percent of our genetic code. Another reason for the gorillas disappearing is habitat destruction through human activities for agricultural activities. This is only going to get worse with the ever-expanding human population.

According to the Endangered Species Act (ESA), a species is listed as endangered only when there is a federal examination of its status and the reasons for its falling population status, (Czech, P., Krausman, P.R., & Devers, P.K. 2000). In addition, there is the investigation of the habitat and recovery planning. Because of the investigation and need for the reasons for their decreasing population there may be more endangered species than are on the list. This means that the endangered species may be much larger than predicted.

Other top endangered species include but not limited to are the giant panda, tigers, whooping crane, blue whale, sea otter, snow leopard, Tasmanian devil, and the orangutan. All these species are on the verge of extinction for many of the same reasons as the elephants and gorillas: poaching, habitat destruction, etc. The giant panda is especially endangered with only 2,500 pandas left in the wild (Britannica). If things go as is, the only creatures of these species will be in zoos, in cages, or not exist at all. This would be a total shameful disgrace because it doesn't need to be this way and is extremely amoral.

What ought to be done is put heavy fines of poachers and instill on people the importance of these animals to the country. If we put in place heavy fines for poaching and importing the profit for poaching will go down. Poachers poach for the profit, so reducing profits will reduce poaching. Another thing we can do is make more sanctuaries and national parks where they can roam free without fear of being shot. Then again, we need to make shooting these animals in parks and sanctuaries strictly prohibited with heavy fines for violators. Habitat destruction is a big key to extinction, so making sanctuaries and national parks are an important step, but there are other things we can do. Like eliminate slash and burn farming. The problem here is that not only does this approach on the lives of wildlife, but also these farms are not always sustainable that leads to more slash and burn. If we keep this up there will be little wilderness left for wildlife.

According to Health in Harmony, there are three main dangers in slash and burn, soil and nutrient depletion, destroying food sources for animals, and respiratory problems and plagues. Initially the ash from burning creates added nutrients, but soon after only a few years leaving the soil depleted of nutrient and not able to sustain vegetable growth and normally leads to more slash and burn, and then more depletion and so on. Slash and burn also destroys biodiversity that wild animals depend on. Then the ash from burning can cause respiratory problems to both animals and human. Ashes can travel for hundreds of miles affecting many, especially the young and elderly. Another factor like slash and burning is logging because it depletes wild life of its natural environment. Clearly, we need to stop slash and burning and start living more sustainably.

According to Foin, F.C., et al. (1998); & Flather, et at (1998), habitat destruction is the largest contributor to species endangerment and extinction is occurring at an unprecedented rate and is due to human activities. Also, according to these authors species endangerment and extinctions has its hot spots that should focus our attention varying on the severity of the areas. According to Czech, B., Krausman, P.R., Dobson, A., Rodriguez, J.P., & Wilcove, D.S. (1997). species list for endangerment are more than one reason, but many factors are leading to their endangerment. The primary causes for habitat depletion are slash and burn, logging and urbanization. This, of course, is partially the result of human's exponential growth in population. If we don't stop over populating like we are there will be eventually little wildlife areas left. That's why we need to start limiting our growth factor. One way we can do this is by trying what China used to do, that is, limiting families to one child. We can do this by limiting support in the form of food stamps, welfare, Medicaid, and college tuition and other aid to families who have more than one child.

CHAPTER IX

OVERPOPULATION

Overpopulation affects everything we just went over and more including: pollution, species endangerment, global warming, feeding, agriculture, forest destruction, loss of diversity, habitat loss, mass extinctions, human health issues, poor nutrition, famines, infant mortality rates, fresh water deuteriation, and over use of limited resources. It effects pollution because humans produce waste. The more there are of us, the more pollution there are going to be from many sources. We already touched on the factory farms. These farms produce more waste and pollution because the animals are overcrowded in small spaces, which is inhumane as well as an environmental hazard. The reason we are using them are to make more money (greed) and to feed our overcrowded planet. Third, countries and especially Asian countries are particularly becoming overpopulated, which has an impact on the rest of the world.

The effects of overpopulation on climate change is multifaceted. First is the over use of automobiles, then there is the increasing use of industry, next is deforestation, slash and burn practice in agriculture and finally there are the factory farms. According to McMichel et al (1999), the effects of population explosion will affect between 10 to 11 million people or more worldwide starting in the 1990s. Also, by McMichel et al (1999), agricultural practices have led to the lowest overall ill health in modern times and some of the highest mortality rates in early life. Also, Malthus, in McMichel et al (1999) predicts that human betterment would fail due to over population will fail unless actions are taken, such as later child bearing and having fewer children. The reasons for human exponential growth are too many children in families, poor reproduction rights, healthcare with lower child mortality, etc. McMichel, et al (1999) have estimated that a population growth between 1985 and 2100 will result in an 35 percent increase in fossil fuel emissions. Also, according to McMichel, et al (1999) most of the environment impact is caused by the wealthiest countries; therefore, it is up to us to initiate environment change and sustainable living.

The Earth's deterioration and expansion in CO_2 emissions per capita from 1880 to 2000 and a projective prediction if we curtail our CO_2 emissions, will exponentially increase global average temperatures McMichel, et al (1999). The CO_2 emissions from 1880 to 2000 went from one ppm (parts per million) to six ppm, an increase of six times in 200 years.

There are numerous threats to the environment associated with over population: urbanization, agriculture, outdoor recreation and tourism, domestic livestock and ranching practices, reservoirs and of water way diversions like dams, pollution of way air and soil, mining, industry and military activities, harvesting of wild species, logging, construction of roads and other infrastructures and drainage of wetlands (Oxford Bioscience). This is a huge amount of damage to our environment due to over population and the population keeps growing.

An urban area is defined as a minimum of 1,000 people residing per square mile with of at least 50,000 people (google). Also, approximately 52 percent of the world's population lives in an urban area, and there are 1,064 major urban areas in the world, with more than 500,000 inhabitants per city (Demographic World Urban Areas, in Newgeography). According to Newgeography, over 50 percent of people live in urban areas from a 100,000 to 10 million. This excessive urbanization is no doubt putting a strain on our environment and encroaching on the habitats of animals–especially upon endangered species worldwide.

As stated earlier to feed the growing population, we need to come up with novel farming practices that was never used before. Cows grazing in fields and chickens running wild have become the exception, not the rule. Because of overpopulation, we no longer have the outdoor space for grazing cattle and other animals. Instead, we keep animals in small cages. According to PETA chickens are the most tortured animal on the planet. Factory farms keep chickens in extremely tight spaces, remove their beaks because in such confined spaces they have an aptitude for pecking at one another. Also, the handling of these birds is despicable. Handlers pick them up by the necks and throw them in cages like they are just a piece of meat and not animals who can feel pain. Chickens like any animal can feel pain. They have a central nervous system and pain sensors throughout their bodies, so of course they feel pain.

Pollution is another effect from overpopulation. Pollution comes primarily in three forms: soil, water, and air. Water pollution is especially severe. For instance, about 80 percent of our stream contain pesticides, pharmaceuticals, and other chemicals. Also, every year, 1.2 trillion gallons of untreated sewage, storm water, and industrial waste are poured into our waters. According to Huamain, C., Chungrong, Z., Cung, T., & Yongguan, Z. (1999), the main source of heavy metals in the soil are irrigation especially from sewage, sludge and mining. Also, according to the same authors sewage have increased in the soil over 300 percent from 1962 to 1993.

Air pollution is also vehement and is partly caused by our ever-growing population. According to NPG.org (Negative Population Growth), electricity generation is our largest source of air pollution contributing to greenhouse gas emissions. Also, according to NPG.org, 2004 air pollution exceeded national health standards as set by and the reported by the U.S. Public Interest and Research Group (PIRG). Furthermore, the Environment Defense Fund have reported that 80 percent of cancer risks are from automobiles pollution. Soil pollution is also a dangerous aspect in our environment. One of the biggest and more detrimental causes of air pollution comes from mining and burning coal. According to Liang, J.Y., C., & Shemg, J.Y, (2003), coal burning is a major source of arsenic in the air, soil, water, and food supply that affects thousands of people worldwide

Coal burning is a major detriment to human health. Not only arsenic either. Coal is a major source of respiratory disease for those living near a coal fired plant. For example, in 1948 in Danora, PA, nearly half of the inhabitants were inflicted with respiratory diseases because of the town's coal fired plant, when atmospheric conditions trapped the toxic gases of coal in the air according to Lockwood, A.H. (2012). More recently in 1990 in Dubliner, Ireland, the town, due to rising oil prices, switched to burning coal to heat their homes.

Despite coal's reputation for bad health there is a great interest in building coal electrical plants. Coal is a major source of power. In the U.S., 45 percent of electricity comes from burning coal and coal fueled the industrial revolution. In 2009, coal reserves reached 908 billion tons. In 2006, the Department of Energy estimated that 53 new coal powered electricity plant will be opened by 2025. Yet, the U.S. Department of Environmental protection listed 67 different air pollutants. So, consider the hazards why do we keep using coal? One answer is that coal is cheap compared to other fuels. Coal costs 10 cents per kwh, oil costs 21.56 per kwh, and gas costs 10 cents per kwh, the same as coal. So, why not use gas that is the same price as coal but comes with much higher health and environmental risks. One reason to keep using coal is that of employment. Coal provided many jobs from miners to transportation to plant workers. Yet, another reason is politics. The United Mines Workers of America union, for instance, has a powerful voice in Washington. No one in Washington wants to put many voting miners out of work and antagonize a powerful union.

Soil pollution is also a big problem and acerbated by overpopulation. According to Huamain, et al. (1999), heavy metals in the soil is prevalent and especially so in China. It is also one of the primary sources of soil pollution worldwide. Heavy metals pollution is defined as where the metals in the soil exceeds the natural amount that is normally in the soil. Heavy metal in the soil has a negative impact on crop production, atmosphere, the water system and human health, according to Huamain, et al (1999). Primary sources of heavy metal pollution are sewage discharge including irrigation, A.H. (pesticide and fertilizers).

Most of all our environmental concerns are either caused by overpopulation or is exacerbated by over population. Why shouldn't this be true? Overpopulation created the need for factory farms, fish farms, the need for pesticides to increase crop output, slash and burn. Factory farms are a direct attempt to feed our ever-growing population. Because we don't have enough land mass to let farm animals run free on farms, we must keep them in very confined spaces in factory farms, and it's the same with fish farms. Pesticides are used to increase farm output because we must be at maximum output on our farms. Then there's greed. The greed to make an easy buck by poaching, for instance. The only environmental problem that is not caused or facilitated by population over expansion is coal burning that is caused by politicians who are afraid to lose the vote of coal miners and others worker within the coal industry. The answer to this dilemma is simple: pay for re-job training. If we pay for different employment training for coal workers these workers won't so vehemently fight for their jobs making it easier to fade out coal. This is a very simple solution and a no brainer.

CONCLUSION

The point of this book is that because nearby intelligent life is rare at best, our planet is special. Because it is special, we should treat it as special and preserve it. We went over all the possibilities of alien life forms visiting us, and I think I made it clear that this is extremely unlikely. I made the argument that considering the plentitude of possibilities life is out there. However, considering the vastness of space, the speed limit within the universe and the difficulties of carrying life supporting material, the chance of extraterrestrial intelligent lifeforms visiting us is very unlikely. Therefore, because of Earth's uniqueness in the nearby universe, this planet is special, and anything special should be taken care of. Also, because there're no nearby planets that are inhabitable, we can't simply destroy this planet and move on to another.

I went through a lot of evidence for UFOs and alien obductions not being ETI, and I think I gave pretty good explanations other than ETI. The evidence can take the form of balloons, aircrafts, planets, temperature inversions, faulty eyewitness accounts, lack of oxygen, the Red Scare, paranoia, false memory syndrome, hypnosis, and the desire to believe (thought is probably unconscious). Add to this is the autonomous difficulties of very long space journeys: the speed of light is unattainable, life support and the vastness of space. Considering the Earthly explanations for UFOs and the difficulties of space travel, I believe I made a pretty good case against ETI visiting us.

We are not alone though, it's just that the vastness of space speed limit and the difficulties of space travel makes it unlikely for ETI visits. I made it perfectly clear that I believe ETI is far away and too difficult for visits. ETI may be a hundred thousand light years away that makes Earth visits next to impossible unless you believe in wormholes, which there is little empirical evidence to support, outside of Einstein's relativity theory. But Einstein's theory is not physical evidence, but just theoretical. Besides, I find wormholes to sound too much like science fiction.

Also, I went through the possibilities of life within our own solar system. The possible life supporting planets and moons are Mars, Europa, and Titan. It is a reasonable possibility that one or more of these words harbor primitive life. However, they don't likely harbor ETI, that is extra-terrestrial intelligences, but more likely primitive life forms, such a bacteria or primitive ocean dwelling creatures. I see it this way because if there were ETI on Europa or Titan, we would be picking up radio or microwave communications from these moons. Furthermore, I question that ETIs could live in a watery world. Then there's Mars, who's only sign of life is nannofossils that could no way be ETI and underground water.

Then I make the argument that because Earth is

unique within a 100,000 or so light years distance, we are unique and should treat our plane as such. I then go into all the evidence for our planet's deterioration, such as climate change, species endangerment, habitat destruction, pollution, factory farms, slash and burn, and overpopulation. I especially focus on global warming because I consider this to be one of our most pressing issues. I talked about the rise in CO_2 and CFCs in raising the global average temperature. I make the distinction between global average temperature and local temperature, which fluctuates due to many variables. However, global average temperature has been regularly rising since the age of industrialization. Overpopulation is behind a lot of our problems including: global warming, species endangerment, pollution, slash and burn, and factory farms.

I also talked about what we can do to save our planet including: green energy such as wind, solar, geo thermal, hydro, less alliance of coal and the carbon tax. On an individual level, we can do more recycling, eat less meat, and be politically active. If we do these things, we can save the planet, but we must act now because the planet is in a dire circumstance.

It is my hope that this book has made it clear that we are not alone but are unique within several thousand of light years. Because of the distance between us and ETI I made a convincing argument for preserving this planet. We must take care of this planet because we are the only one like it within several hundreds or thousands of light years distance.

BIBLIOGRAPHY

Anomalies. anomalies.net/archive/Text-Archive/txt3/2401.ufo

Archaeologyinfo: http://www.archaeologyinfo.com/homoerectus.htm

BB Archives: prejectbluebookarchive, org; UFO Casebook

Berkeley: https://evolution.berkeley.edu/evolibrary/article/speciationmodes_05

Breasted, J.H. Ancient Times: A History of the Early World; An Introduction to the Study of Ancient (1914) History and the Career of Early Man. Outlines of European History 1. Boston: Ginn and Company, 1914, p.85

Britannica. Falling Stars: 10 of the most famous endangered species. https britannica.com/list/10-of-the-most-famous-endangered-species

Britannica Online Encyclopedia: https://www.britannica.com/science/chlorofluorocarbon

Carbon Tax Center: https://www.carbontax.org/whats-a-carbon-tax/

Center for UFOs http://www.cufos.org/Roswell_fs1.html

Chen Huamain, Zheng Chunrong, Tu Cong and Zhu Yongguan

Ambio Vol. 28, No. 2 (Mar., 1999), pp. 130-134
Heavy Metal Pollution in Soils in China: Status and Countermeasures Published by: Springer on behalf of Royal Swedish Academy of Sciences

Compassion in World Farming, https://animalcharityevaluators.org/charity-review/compassion-in-world-farming-usa-ciwf/

https://www.ciwf.org.uk/factory-farming/environmental-damage/?gclid=EAIaIQobChMI6vrxtc3d3QIVSwOGCh16-QHPEAAYASAAEgJ64fD BwE#resourcewaste

Chalker, B. (2005). Hair of the Alien: DNA and other Forensic Evidence of Alien Abductions.

Clark, J. (1997). *The UFO Book: Encyclopedia of the Extraterrestrial.* Visible Ink,. ISBN 1578590299.

Crystallinks. ://www.crystalinks.com/ancientaircraft.html

Cornell University. http://curious.astro.cornell.edu/about-us/77-the-universe/extrasolar-planets/general-questions/323-how-are-planets-detected-around-other-stars-intermediate

Czech, B., Krausman, P.R., & Devers, P.K.(2000) Economic Associations among Causes of Species Endangerment in the United States: Associations among causes of species endangerment in the United States reflect the integration of economic sectors, supporting the theory and evidence that economic growth proceeds at the competitive exclusion of nonhuman species in the aggregate. *BioScience*, 50, (7, 1), 593–601.

Darwin, C. (1999). The Origin of Species

Dawkins, R. & Wong, Y. (2005). The Ancestor's Tale: A Pilgrimage to the Dawn of Evolution

Dummies.com http://www.dummies.com/how-to/content/different-patterns-of-evolution.html

Encyclopedia of Science: http://www.daviddarling.info/encyclopedia/M/Mantell.html

Enviropedia: http://www.enviropedia.org.uk/Global_Warming/CFCs.php

Exoplanet: http://exoplanet.eu/catalog.php

False Memory Syndrome Foundation. http://www.fmsfonline.org/

Flather, C.H., Knoles, M.S., & Kandall, I.A. (1998). Threaten and endangered species geography: characteristics of hot spots in the conterminous United State1s. Bioscience, 48 (5)

Foin, C.H., Knowles, M.S., & Kendall, I.A. (1998). Improving recovery planning for threatened and endangered species: Comparative analysis of recovery plans can contribute to more effective recovery planning. Bioscience, 48 (2).

Health in Harmony. https://healthinharmony.org/

Huamain, C., Chunrong, Z.,& Yungguan, T.C.Z. Heavy metals polutions in soild in China status and countermeasures. Royal Swedish Academy of Science. 28 (2), 130-134.

Idealist finder. Ballpoint Pen: http://www.ideafinder.com/history/inventions/ballpen.htm

Iowa State University: https://www.abe.iastate.edu/extension-and-outreach/carbon-monoxide-poisoning-vehicles-aen-208/

Kayla Garnet Rose, youtube

Kenneth, A. : http://www.project1947.com/fig/ka.htm

Kaplan, R. & Manicavasagar, V. (2001). Is there a false memory syndrome? A review of three cases. Comprehensive Psychiatry, 42 (4), 342-348.

Kepler: kepler.nasa.gove/

Liang, J.Y., & Cao, C.H.G. <u>& Sheng, W.Y. C. (2003). Study of distribution of endemic arsenic in China. Journal of Hygiene Research</u>, 32 (6):519-540]

Loftus, E, & Zanni, G. (1975). The wording of a question: The influence of the wording of a question. Bulletin of the Psychonomic Society.

Lockwood, A.H. (2012) The silent epidemic: coal and the hidden threats to health:

Loftus, Elizabeth, F. (1975). Eyewitness testimony: The influence of the wording of a question. Bulletin of the Psychonomic Society, 5 (1) 86-88.

McMichel, <u>A. H</u> (1990) professor of epidemiology and <u>J W Powles</u>, US National Library of Medicine, National Institute of health.

McMichel, A.H., & Powles, J.W. (1999) Human numbers, environment, sustainability and health. U S Library of Medicine National Institute of Health. 319, (7215), 977-980. https://www.ncbi.nlm.nih.gov/pmc/articles/PMC1116806/

Moss, A.R., Jouany, J.P., & Newbold, J. (1999). Methane production by ruminants: Its contribution to global warming. Rowett Rearch Institute.

Mudd, Kimberly & Bovern, John M. (2004). Conformity Misinformation and time delay negatively affect eyewitness confidence and accuracy. North American journal of Psychology, 6, (2) 227-238.

*NASA.gov: https://asd.gsfc.nasa.gov/blueshift/index.php/2015/07/22/how-many-stars-in-the-milky-way/
*NASA.gov:https://www.nasa.gov/feature/goddard/2018/hubble-uncovers-the-farthest-star-ever-seen
*NASA.gov press release: https://www.nasa.gov/press-release/nasa-confirms-evidence-that-liquid-water-flows-on-today-s-mars
NASA: Global Climate Change Vital Signs of the Planet: https://climate.nasa.gov/news/2811/2018-arctic-summertime-sea-ice-minimum-extent-tied-for-sixth-lowest-on-record/
*NASA: Planet-Hunter Kepler telescope explained. https://www.space.com/17383-kepler-planet-hunting-nasa-telescope-infographic.html
*NASA.gov Marson water: https://www.nasa.gov/press-release/nasa-confirms-evidence-that-liquid-water-flows-on-today-s-mars
*NASA.gov.: https://spaceplace.nasa.gov/review/dr-marc-space/solar-systems-in-galaxy.html
*NASA; Space.com
*NASA, TESS: https://www.wired.com/story/nasas-new-exoplanet-satellite-has-a-better-shot-of-finding-life-close-to-home/
NICAP doc: http://www.nicap.org/docs/SignRptFeb1949.pdf
NSIDC: National Snow & Ice Data Center, http://nsidc.org/arcticseaicenews/2018/09/arctic-sea-ice-extent-arrives-at-its-minimum
Newgeography.com
NICAP WNS: http://www.nicap.org/wns.htm
Oparin, A, I. (1953). The Origin of Life.

Oxford, Bioscience: https://academic.oup.com/bioscience/article/50/7/593/354580

Parkinson, C.L., & DiGiolamo, N.E. (2016). New visualizations highlights new information on the contrasting Arctic and Antarctic sea-ice trends since the 1970s. Remote Sensing of the Environment, 183, 198-204.

Pearce. D. (1991). The role of carbon taxes in adjusting to global warming. The Economic Journal. 1, 938-948. https://www.jstor.org/stable/2233865?newaccount=true&read-now=1&seq=1#metadata_info_tab_contents

Perry, C. & Gold, A.D. (Nov. 1995). Hypnosis and the elicitation of true and false memories of childhood sexual abuse. Psychiatry, Psychology and Law, 2, 127-138.

Physicsforums http://www.physicsforums.com/showthread.php?t=156291

PIRG: https://uspirg.org/sites/pirg/files/cpn/USN-100518-A1/index.html

Planet Quest: http://planetquest.jpl.nasa.gov/atlas/atlas_index.cfm

http://exoplanet.eu/catalog.php

Project Blue Book. http://www.ufoevidence.org/topics/projectbluebook.htm

Project 47: (www.project1947.com/fig/ka.htm)

Ufo.about: http://ufos.about.com/od/aliensalienabduction/p/andreasson.htm

Quest.nasa: quest.nasa.gov/sso/cool/pioneer10/mission/

Rense. http://www.rense.com/general7/ages.htm

Roswell prove: http://roswellproof.homestead.com/Marcel_evaluations.html

Science Daily: http://www.sciencedaily.com/releases/2009/01/090109173205.htm

Schneider, T., DR. How we know global warming is real: The science behind human-induced climate change. Skeptic.com

Science Daily; https://www.sciencedaily.com/releases/2009/01/090109173205.htm

Scribd: http://www.scribd.com/doc/92445/Primordial-Soup-Theory

SETI.org

Titan; https://solarsystem.nasa.gov/moons/saturn-moons/titan/in-depth/

UFO Casebook: http://www.ufocasebook.com/1951fortmonmouth.html

Ultra Hypnosis, youtube

Volcanoes National Park http://volcanoesparkrwanda.org/information/people-poach-gorillas/

Wempon, K. https://newsnetwork.mayoclinic.org/discussion/are-you-getting-too-much-protein/

Washingtonpost.com (May 30, 2005)

Waterman, L., & Purvis, W. Fountain Pen. http://inventors.about.com/library/weekly/aa100897.htm

Weatherquestions: http://www.weatherquestions.com/What_is_a_temperature_inversion.htm

What really happened at Roswell: haphttps://www.csicop.org/si/show/what_really_happened_at_roswell

Whipnet.org. http://ufo.whipnet.org/xdocs/alexander.the.great/index.html

Wikipedia, http://en.Wikipedia.org/wiki/1952_Washington_D.C._UFO_incident

Wikipedia handicap: https://en.Wikipedia.org/wiki/Handicap_principle
https://en.Wikipedia.org/wiki/Project_Blue_Book
Wikipedia. Project Grudge: https://en.Wikipedia.org/wiki/Project_Grudge
Wikipedia Mantell UFO incident: https://en.Wikipedia.org/wiki/Mantell_UFO_incident
World Wildlife Fund (WWF)

www.ingramcontent.com/pod-product-compliance
Lightning Source LLC
Chambersburg PA
CBHW020434220526
45464CB00002B/698